HOW DARK MATTER CREATED DARK ENERGY AND THE SUN

An Astrophysics Detective Story

JEROME DREXLER

Universal Publishers
USA • 2004

How Dark Matter Created Dark Energy and the Sun:
An Astrophysics Detective Story

Copyright © 2003 Jerome Drexler
All rights reserved.

This book is protected by copyright. No part of it may be reproduced or translated without the prior written permission of the copyright owner, except as permitted by law.

Universal Publishers/uPUBLISH.com
USA • 2004

ISBN: 1-58112-551-8

BNT 15.95 9/05

http://www.uPUBLISH.com/books/drexler.htm

This book is dedicated to Sylvia, my wife and lifelong partner. All royalties from this first edition are being paid directly to one of her favorite charitable organizations, Recording for the Blind & Dyslexic, Palo Alto, California.

CONTENTS

PREFACE	x
INTRODUCTION: How Dark Matter Created Dark Energy and the Sun – An Astrophysics Detective Story	1
Can the Ultra-High-Energy (UHE) Proton Be a Dark Matter Candidate?	1
Could Cosmic-Ray Protons Play a Leading Role in Creating the Sun?	2
UHE Protons and Cosmic-Ray Protons Exhibit a Characteristic Normally Expected of Dark Matter Particles	3
Halo UHE Protons Seem to Be the Source of Galaxy Cosmic-Ray Protons	3
What Causes the Transformation From UHE Protons to Cosmic-Ray Protons?	4
The Accelerating Expansion of the Universe Is Between Galaxy Clusters	4
The Focus of the Lecture Slides	6
LECTURE: "How Dark Matter Created Dark Energy and the Sun"	8
PART I: The Discovery and Mystery of Dark Matter	8

Discovery and Confirmation That a Dark Matter Halo Surrounds Spiral Galaxies and Clusters	9
Dark Matter – Astronomers Cannot See It With Telescopes	10
The Massive Dark Matter Halo of a Spiral Galaxy	11
Five Times as Much Dark Matter as Galaxy Matter	12
Is a Universe Made Only of Baryons Impossible?	13
The Search for Dark Matter – It Cannot Be Protons	14
Cold and Warm Non-Baryonic Dark Matter Are Generally Accepted	17
Composition of the Universe – Baryonic Matter, Dark Matter, Dark Energy	18
PART II: UHE Protons as a Dark Matter Candidate	19
Are Protons Near the Speed of Light, Dark Matter Candidates?	20
Are Relativistic Protons Dark Matter Particles?	21
Energy of Relativistic Protons and Their Relativistic Mass	22
Highly Energetic Protons Striking the Earth as Cosmic Rays	23
Cosmic-Ray Energy Distribution at the Earth	24

Evidence That Relativistic Halo Protons Could Be the Long-Sought Dark Matter	25
Halo Protons Create Beryllium and Boron – Additional Dark Matter Evidence	28
The Source of Cosmic-Ray Protons Surrounds the Earth – More Dark Matter Evidence	29
Similar Ratio of Protons to Helium Nuclei in Cosmic Rays and in the Universe	30
PART III: The Dark Matter Halo of Spiral Galaxies	31
Linking UHE Protons and Cosmic-Ray Protons to Dark Matter Particles	32
Masses of Hot "Missing" Intergalactic Baryons Could Be Dark Matter Particles	36
Astronomical Evidence Supports UHE Protons as Dark Matter Particles	37
Particle Physics Evidence Supports UHE Protons as Dark Matter Particles	38
PART IV: Are UHE Dark Matter Halo Protons Relics of the Big Bang?	39
Did All Protons Convert to Hydrogen 700,000 Years After the Big Bang?	43
Big Bang Proton Energies Have Declined to One Billionth of Initial Levels	44

How Dark Matter Created Dark Energy and the Sun   vii

PART V:	The UHE Protons in the Halos of Spiral Galaxies Still Retain Enormous Energies	45

Synchrotron-Radiation Energy Loss of Protons Over 13.7 Billion Years   46

PART VI:	UHE Protons Create Synchrotron Radiation in the Form of Gamma Rays	48

Gamma-Ray Glow Bathes Milky Way   49

X Ray Synchrotron Radiation From UHE Electron Cosmic Rays   52

PART VII:	Big Bang Origin of 10^{20} eV Cosmic-Ray Protons Found in the Milky Way?	53

PART VIII:	The Accelerating Expansion of the Universe and Dark Energy	56

Points Used to Explain Dark Matter and the Accelerating Expansion of the Universe   62

The Accelerating Expansion Between Galaxy Clusters   65

Jerome Drexler's Theory of the Accelerating Expansion Between Galaxy Clusters   66

PART IX:	Jerome Drexler's Theory of Star and Sun Formation	69

What Source of Hydrogen Created the Stars of the Milky Way?   70

About 1 to 2 Pounds Per Day of Cosmic-Ray Nuclei Arrive at the Earth	72
The Sun Came Into Being 9 Billion Years After the Big Bang	73
About 2×10^{18} Pounds Per Day of Cosmic-Ray Particles Strike the Solar System	74
Maybe UHE Cosmic-Ray Nuclei Did Create the Sun	75
From the Sun to Population I Stars in Other Spiral Galaxies	76
The Sun's Mass May Be Greater Today Than at Birth	77
Did Cosmic-Ray Nuclei Trigger the Sun's Fusion Reaction?	78
PART X: Cosmic-Ray Cosmology: Drexler's Unified Theory of Dark Matter, Accelerating Expansion, and Star Formation	79
PART XI: Drexler's Theory of "Immortal" UHE Protons, "Mortal" Cosmic-Ray Protons, and the "Death Spiral"	84
What is the Difference Between a UHE Proton and a Cosmic-Ray Proton?	85
From "Immortal" UHE Protons Into "Mortal" Cosmic-Ray Protons Via the "Death Spiral"	86
PART XII: Astronomers Report Elliptical Galaxies With No Dark Matter Halo	89

PART XIII: Cosmic-Ray Cosmology Applied to Galaxy Formation	91
Galaxy Formation – The Proton Larmor Radius	92
Only One Dark Matter Candidate Establishes the Approximate Size of the Milky Way	94
Galaxy Formation – Some Plausible Speculations	95
PART XIV: The Principal Goal of These Lectures Is to Provide Evidence	101
PART XV: Tentative Conclusions Regarding Dark Matter, Accelerating Expansion, Star and Sun Formation, and General Astrophysics Theory	104
Tentative Conclusions: Dark Matter	105
Tentative Conclusions: Accelerating Expansion of the Universe Between Galaxy Clusters	107
Tentative Conclusions: Star and Sun Formation	108
Tentative Conclusions: Galaxy Formation	109
Tentative Conclusions: General Astrophysics Theory	110
CONCLUSION	113
APPENDIX: A Dozen Contrarian Astrophysics Ideas	114
BIBLIOGRAPHY AND SUGGESTED SOURCES	116
GLOSSARY	119
INDEX	135

PREFACE

Through use of a lecture-slide format, this book presents an astrophysics detective story. It chronicles my search for astronomical clues and evidence to unveil the nature of dark matter. My original goal was to identify dark matter, a decades old mystery. In the process, I developed a new theory for dark matter and believe I have illuminated the nature of dark energy and the process of Sun formation. A unified theory for all three phenomena evolved.

This lecture material was originally prepared as a 30-slide PowerPoint lecture that I had planned to present at a university colloquium. As I proceeded to develop the material, it grew to over 100 presentation slides representing about three lectures. Also, more and more original ideas permeated the material, to the point that the lectures began to look like technical papers on three different subjects – dark matter, the accelerating expansion of the Universe (dark energy), and star (and Sun) formation.

At that point, I decided that the best way of disseminating the information would be to publish the three lectures/technical papers both online and in paperback book form. By these procedures, the lectures could be read by a wider audience.

These lectures are directed toward researchers, professors, postdocs, graduate students, and individuals who majored as undergraduates in astrophysics,

physics, cosmology, astronomy, or related fields. A 90-word glossary is also provided.

My interest in and knowledge of particle physics began early in my professional career. Many years ago at Bell Laboratories in Murray Hill, New Jersey, my seven years of research and development were related to the interaction between microwave electromagnetic waves and electron beams and streams that caused the electrons within them to be bunched, focused, modulated, or deflected. I spent a second seven years working in the same technologies at a company called S-F-D Laboratories, Inc.* that I co-founded in New Jersey with my Bell Labs associates, Dr. Joseph A. Saloom and Dr. Joseph Feinstein. Those 14 years of R&D made me knowledgeable of the behavior of high-speed charged particles under the influence of static magnetic and electric fields and electromagnetic microwaves in a vacuum system, not unlike the behavior of charged particles in the vacuum of outer space.

It was with this background that I began my weekend literature studies of astrophysics, cosmology, and dark matter in 1994 through the analysis of astronomical data and writings. Following the announcements of the accelerating expansion of the Universe and dark energy in 1998 and 1999, those phenomena were added to my studies. By 2001, my focus was on ultra-high-energy

---

*Multiple SFD 262, SFD 263, or SFD 268 crossed-field amplifiers are used as the microwave power sources in phased-array radars for U.S. Navy Aegis destroyers and cruisers.

(UHE) protons as being the key dark matter constituent. Note that when UHE protons rain on the Earth's atmosphere, we call them cosmic-ray protons, which have been observed by astronomers for about 90 years. By 2002, my research was expanded to seek any possible astronomical or theoretical links between dark matter and dark energy.

The theories developed and conclusions drawn in this book are based upon analyses of astronomical data published by astronomers, astrophysicists, and cosmologists in books, scientific papers, and other literature. Any astronomical data mentioned in this book would have been derived from those other sources, which in most cases are identified. This book is offered as a thought-provoking exploration and analysis and to encourage discussion and debate.

HOW DARK MATTER CREATED DARK ENERGY AND THE SUN

INTRODUCTION

Can the Ultra-High-Energy (UHE) Proton Be a Dark Matter Candidate?

For the past 15 years or more, the world's top astrophysicists and cosmologists have been claiming for various reasons that baryons (protons or neutrons) could not be the principal constituent of dark matter. Unfortunately, my favorite dark matter candidate, ultra-high-energy (UHE) protons, happens to be a baryon. Some astrophysicists have been proposing the WIMP (weakly interacting massive particle), a heavy theoretical non-baryonic particle, as their favorite candidate, which has some similarities to the neutralino, a heavy theoretical non-baryonic particle, being proposed today.

The UHE proton as a dark matter candidate has several key advantages over the theoretical WIMP and neutralino particles (which have never been detected). The cosmic-ray protons are real and have been detected by astronomers for about 90 years. Further, a cosmic-ray proton has an enormous kinetic energy ranging from 10^{10} electron volts to 10^{20} electron volts and therefore has an enormous relativistic mass and an enormous gravitational attraction – orders of magnitude greater than those of the theoretical WIMP and neutralino.

The UHE proton is also difficult to detect by astronomical methods – a necessary characteristic for dark matter particles.

Could Cosmic-Ray Protons Play a Leading Role in Creating the Sun?

If cosmic-ray protons are a principal constituent of dark matter, they might have been involved in creating the Milky Way, its stars, and our Sun. In late 2002, I tested that concept by using a comment in Herbert Friedman's famous 1998 book entitled, *The Astronomer's Universe*. Friedman wrote, "Eventually, balloon-borne experiments proved that cosmic rays are charged particles consisting mostly of hydrogen nuclei (protons), helium nuclei (alpha particles), a few heavier nuclei, and electrons. All told, about 1 - 2 pounds per day arrive at the earth."

It occurred to me that the density of 1 to 2 pounds per day onto the Earth might be representative of the cosmic-ray particle density raining onto the entire solar system region during the last 13.5 billion years. Could those cosmic-ray protons and other cosmic particles have added any significant mass to the Sun? I did some math and discovered that the cumulative cosmic-ray particle mass showering the solar system region over 13.5 billion years would be in the range of about 3×10^{30} to 6×10^{30} kilograms. I was very encouraged when I compared that accumulated cosmic-ray particle mass to astronomers' standard estimate of the mass of the Sun at 2×10^{30} kilograms. Not only are the two mass quantities close, but the Sun's mass is lower, which is a more plausible expectation.

UHE Protons and Cosmic-Ray Protons Exhibit a Characteristic Normally Expected of Dark Matter Particles.

Many cosmologists, and I, have believed that dark matter was somehow utilized in creating the galaxies and the stars. Now, there is some circumstantial evidence that the mass of the Sun may have been derived from the mass of cosmic-ray protons and helium nuclei raining upon the solar system for 9 billion years originally, and for another 4.5 billion years since the birth of the Sun. This seems to provide support that the cosmic-ray protons may have been instrumental in creating the Sun; and, if true, that cosmic-ray protons and their source, UHE protons, may be the long-sought dark matter particles.

Halo UHE Protons Seem to Be the Source of Galaxy Cosmic-Ray Protons.

However, the long-sought dark matter particles seem to reside in a thick halo around spiral galaxies and galaxy clusters. If UHE cosmic-ray protons are dark matter particles, what would keep them in the halo of galaxies and galaxy clusters? My answer is that the extragalactic magnetic field strength, which is very weak but has just the right strength to keep the positively charged very high speed UHE protons in circular and spiral orbits around galaxies. The dark matter halo has a thickness which is probably created by UHE protons having various kinetic energies and therefore different spiral path diameters in the extragalactic magnetic field of the galaxy halo.

What Causes the Transformation from UHE Protons to Cosmic-Ray Protons?

The next question is, why do the UHE protons fall to Earth as cosmic-ray protons? The following lectures explain how the UHE protons moving in the extra-galactic magnetic field lose energy slowly through what is called synchrotron radiation, which causes them to slow down. This reduces the size and the radius of curvature of their spiral orbits, causing the synchrotron radiation to increase further. Finally, when UHE protons encounter the higher magnetic field of the Milky Way galaxy, their synchrotron radiation losses skyrocket and the protons plunge into the galaxy as cosmic-ray protons, in what I call a "death spiral."

The Accelerating Expansion of the Universe Is Between Galaxy Clusters.

Another significant step in developing the basis for the lectures was achieved when I developed a theory to explain the accelerating expansion of the Universe utilizing relativistic UHE protons, the galactic and extragalactic magnetic fields, and two principles of physics.

In 1929, Edwin P. Hubble announced that with the exception of the galaxies closest to the Milky Way, galaxies are rushing away from each other in all directions and, therefore, the Universe is expanding. Maybe the word "expanding" should have been in quotes since the solar system is not expanding, the Milky Way galaxy is not expanding, and in the Local

Group of galaxies, the Milky Way and the Andromeda galaxy are actually moving toward each other.

That is, Hubble's expanding Universe discovery is that, at the greatest distances, galaxy clusters are all moving away from one another. For many years, it was not realized that the expansion was accelerating. Analyses of ancient supernova explosions in distant galaxies during the period 1998-2003 disclosed that the separation velocities between galaxy clusters have been increasing, making the Universe expand faster and faster. This accelerating expansion phase apparently began about 5 billion years ago.

In searching for an explanation for the accelerating expansion of the Universe, I felt the most significant clue I had was that it only applies to the separation of galaxy clusters and not to the separation of galaxies or stars within a galaxy. This sent me a message that the sought-after explanation seemed to be associated with some phenomenon, rather than a field theory like antigravity. Another important clue that helped me confirm the theory I had already developed, was that the expansion acceleration began about 5 billion years ago.

When I began focusing on the mystery of accelerating expansion, I already had established in my mind the strong dark-matter candidacy of the relativistic UHE proton, which could be the principal constituent of the dark matter halos around spiral galaxies and galaxy clusters. Could the relativistic UHE protons help provide a plausible explanation for the accelerating expansion?

Having made the mental connection between galaxy clusters and their halos comprising UHE proton dark matter, it occurred to me, for several reasons, that two well-established principles of physics might lead me to a plausible explanation for the accelerating expansion of the separation between galaxy clusters. They are:

(1) When relativistic protons pass through the extra-galactic or galactic magnetic fields, they are deflected into spiral paths, causing them to generate synchrotron radiation which in turn causes them to lose kinetic energy and to thereby lose relativistic mass; and

(2) The linear momentum of an object equals the product of its mass and its velocity. If no external forces are acting on a group of galaxy clusters, the Law of Conservation of Linear Momentum requires that the total linear momentum of all the galaxy clusters in the group shall remain constant.

In the lectures, these concepts are developed into an explanation and theory for the accelerating expansion of the Universe.

The Focus of the Lecture Slides.

Four principal points are the focus of the lecture slides comprising this manuscript:

(1) Dark matter is simply ultra-high-energy (UHE) protons circulating around galaxies and galaxy clusters in a very thick halo.

(2) The minuscule magnetic fields of the extragalactic Universe and the slightly higher magnetic fields of galaxies may be small, but they control the spiral paths and life spans of the powerful and multitudinous UHE protons.

(3) The accelerating expansion of the Universe probably comes about by the continual loss of relativistic mass of the dark matter halos around galaxy clusters, which reduces their gravitational attraction and also causes those clusters to automatically speed up to maintain their linear momentum under the Law of Conservation of Linear Momentum.

(4) It appears that the Sun may have been created by UHE protons that had been circulating in the dark matter halo around the Milky Way but had plunged into the Milky Way in the form of cosmic-ray protons after losing a very large percentage of their energy due to synchrotron radiation.

In the following slide-format lectures, the explanations for dark matter, accelerating expansion of the Universe (which some cosmologists attribute to dark energy), and star (and Sun) formation are all based upon a unified theory which involves UHE protons traveling through at least two levels of weak magnetic field strengths. Only 20th century physics is employed, and only known types of particles are utilized.

With that said, let the lectures begin.

LECTURE:

HOW DARK MATTER CREATED DARK ENERGY AND THE SUN

PART I

THE DISCOVERY AND MYSTERY OF DARK MATTER

SLIDE #1

Dr. Fritz Zwicky Discovered and Dr. Vera Rubin Confirmed That A Dark Matter Halo Surrounds Spiral Galaxies and Clusters

- In 1933, Fritz Zwicky, an astrophysicist at the California Institute of Technology, was studying a nearby cluster of galaxies called the Coma cluster. He observed some galaxies in the cluster traveling at unexpectedly high speeds. Since the observable mass in the cluster was not sufficient to provide the gravity needed to hold galaxies moving at such high speed, he felt there must be a great halo of invisible mass surrounding the cluster.

- In 1977, Vera Rubin and her associates at the Carnegie Institution studied the rotation curves of galaxies by measuring the Doppler velocities of stars in various locations within 60 spiral galaxies. The rotation curves for the stars were nearly flat, indicating that a halo of invisible mass must surround each galaxy.

SLIDE #2

Dark Matter – Astronomers Cannot See Dark Matter With Their Telescopes

- Despite the absence of telescopic evidence, astronomers believe in the presence of dark matter because they detect its gravitational influence or because certain theories predict its existence. They believe that the halo of spiral galaxies harbors dark matter because they notice its gravitational influence on the stars they can see.

- Inflationary cosmologists, who believe the Universe will continue to expand, also believe that the Universe is full of dark matter because inflation theory predicts that the Universe has a large mass density.

- Inflationary cosmologists believe that particle synthesis during the Big Bang would not create enough protons and helium nuclei to achieve the large mass density necessary for continued expansion of the Universe.

SLIDE #3

The Dark Matter Halo Is the Massive Outer Region Of a Spiral Galaxy

- The dark matter halo is the massive outer region of the Milky Way that surrounds the disk and stellar halo. The dark matter halo consists mostly of dark matter particles whose form has been unknown. Though it emits almost no light, the dark matter outweighs the ordinary matter in the galaxy by a factor of about five.

- Studies in 1999 found that the dark matter halo of a spiral galaxy extends much farther into space than previously estimated, extending 10 to 20 times the size of the visible regions of the galaxies rather than 4 to 6 times.

SLIDE #4

There Is About Five Times as Much Dark Matter as Ordinary Baryonic Galaxy Matter

Many cosmologists believe that the ordinary baryonic matter (protons and neutrons) of stars, galaxies, and gas accounts for no more than 5% of the mass of the Universe. They believe the invisible or dark matter that has so far eluded detection by the best instruments of astronomy or particle physics accounts for about 25% of the mass of the Universe and that dark energy, which causes the expansion of the Universe to accelerate, accounts for the remainder. (See slide #11.)

SLIDE #5

"It Would Be Impossible to Build a Universe Made Only Out of Baryons," States MIT's Dr. Alan H. Guth

- One reason that dark matter is required by cosmologists is that ordinary matter, known collectively as baryons, would not have coalesced quickly enough around the embryonic mass fluctuations, according to Dr. Alan H. Guth, an astrophysicist at the Massachusetts Institute of Technology. Ordinary matter would not have immediately been attracted to the weak gravity of the primordial ripples; certain kinds of dark matter could have been.

- "Even before COBE [Cosmic Background Explorer] there were strong suggestions that it would be impossible to build a Universe made only out of baryons," Dr. Guth said, and "COBE helps nail down that conclusion."

SLIDE #6

"It [Dark Matter] Cannot Be Protons Or Neutrons..." States UCLA Professor David B. Cline in "The Search for Dark Matter"

UCLA Professor of Physics and Astrophysics David B. Cline said:

"What kind of particle could dark matter be made of? Astronomical observation and theory provide some clues."

- "It cannot be protons or neutrons or anything that was once made of protons or neutrons."

- "According to calculations of particle synthesis during the Big Bang, such particles were simply too few in number to make up the dark matter."

- "Those calculations have been corroborated by measurements of primordial hydrogen, helium, and lithium in the Universe."

SLIDE #7

"The Search for Dark Matter" By Professor David B. Cline; March 2003 Issue of *Scientific American*

According to Cline's paper:

- The best fit to astronomical observations is "cold" dark matter.

- The current Standard Model of elementary particles contains no examples of particles that could serve as dark matter.

- Of all the extensions of the Standard Model, the theoretical non-baryonic neutralino has gotten the most attention recently.

- The theoretical neutralino is "an amalgam of the super-partners of the photon, the Z boson and perhaps other particle types." Although the neutralino is heavy, it is the lightest super-symmetric particle and therefore should be stable since it cannot decay into a lighter version.

"The Search for Dark Matter" By Professor David B. Cline; March 2003 Issue of *Scientific American*

- In Cline's graph on page 59 of the *Scientific American* article, the sought-after dark matter neutralino particles are predicted to have a mass in the range of about 35 to 10,000 times greater than the mass of a proton.

- At first glance, the graph seems to rule out proton-based dark matter since a proton's rest mass is too low.

- However, protons traveling near the speed of light with kinetic energies ranging between 3.5×10^{10} eV to 1×10^{13} eV would have a relativistic mass of about 35 to 10,000 times the mass of a proton – the same as the theoretical neutralino.

SLIDE #9

Cold and Warm Dark Matter Have Been Winning the Race for Inclusion in the Dark Matter Scientific Paradigm

- Professor Cline's table on page 53 of the *Scientific American* article shows cold dark matter at 25% and dark energy at 70% of "Probable Contribution to Mass of Universe." (See slide #11.)

- Cold (and warm) dark matter consists of hypothetical non-baryonic particles that move slowly.

- Hot dark matter consists of hypothetical non-baryonic particles that move rapidly.

- When cosmologists run computer simulations of the formation of galaxies based upon the mass density ripples of the very early Universe, they find that cold dark matter seems to work better on a large scale and hot dark matter works better on a small scale. They have leaned toward cold dark matter.*

*Computer simulations of the formation of the Universe using cold dark matter led to too many dwarf galaxies, so some computer simulations have utilized warm dark matter.

SLIDE #10

COMPOSITION OF THE UNIVERSE

(Derived from table on page 53 of the March 2003 Issue of *Scientific American*)

Material	Representative Particles	Probable Contribution To Mass of Universe	Sample Evidence
Ordinary ("baryonic") matter	Protons, electrons	5%	Direct observation, inference from element abundances
Cold dark matter	Super-symmetric particles?	25%	Inference from galaxy dynamics
Dark energy	"Scalar" particles?	70%	Supernova observations of accelerated cosmic expansion

SLIDE #11

PART II

ULTRA-HIGH-ENERGY (UHE) PROTONS AS A DARK MATTER CANDIDATE

SLIDE #12

Are Protons That Are Traveling Close to the Speed of Light, Dark Matter Candidates?

- A key part of the scientific paradigm for dark matter today involves negative statements; that is, "It would be impossible to build a universe made only of baryons," [Guth] and "It [dark matter] cannot be protons or neutrons." [Cline]

- However, if the mass of relativistic protons was at least 35 times greater than the rest mass, the above-indicated part of the scientific paradigm probably would not have excluded relativistic protons as dark matter candidates since they would have sufficient mass. [Drexler]

- Thus, if galaxy halo particles were primarily relativistic protons with relativistic masses at least 35 times the proton rest mass, they could be considered to be dark matter candidates. [Drexler]

SLIDE #13

Would Relativistic Protons Fall Within The Gray Area of Cline's Graph of Sought-After Dark Matter Particles?

The answer is *yes*. Cosmic-ray protons that are traveling near the velocity of light, some of which have kinetic energies ranging between 3.5×10^{10} and 10^{13} electron volts and therefore have relativistic masses covering the range of 35 to 10,000 times the mass of a proton (see slide #9), seem to fit Professor Cline's dark-matter-particle mass requirements. [Drexler]

SLIDE #14

Energies of Relativistic Protons Versus Their Relativistic Mass — Where Do They Exist in Nature?

Energy 9.38 x 10^8 eV	Relativistic Mass of a Proton In Terms of Its Rest Mass, m_o
10^{10} eV	11 m_o
10^{11} eV	110 m_o
10^{12} eV	1,100 m_o
10^{13} eV	11,000 m_o
10^{14} eV	110,000 m_o
10^{15} eV	1,100,000 m_o
10^{16} eV	11,000,000 m_o
10^{17} eV	110,000,000 m_o
10^{18} eV	1,100,000,000 m_o

$$\text{Relativistic Mass} = \frac{\text{Energy (in joules)}}{C^2 \text{ (in meters/sec)}}$$

SLIDE #15

Such Highly Energetic Protons Can Be Found Striking the Earth's Atmosphere As Cosmic-Ray Protons

Approximate Kinetic Energy	Approximate Cosmic Ray Flux On the Earth's Atmosphere
10^8 to 10^{10} eV	Slightly less than 1,000 particles per square meter per second
10^{11} eV	One particle per square meter per second
7×10^{15} eV	One particle per square meter per year
3×10^{18} eV	One particle per square kilometer per year
10^{19} eV	3 to 4 particles per square kilometer per century

SLIDE #16

Cosmic-Ray Energy Distribution At the Earth*

CERN Courier, Vol. 35, No. 10, December 1999

[See Slide #16 for the Key Data Points]

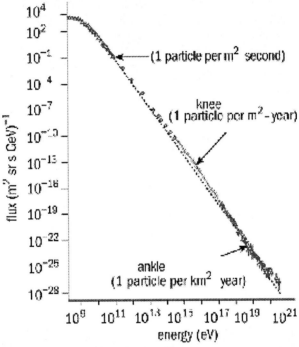

"The cosmic-ray energy distribution shows remarkable uniformity over 10 orders of magnitude. However, there are two kinds. The ACCESS experiment is designed to investigate 'the knee' (near 10^{15} eV)."

*Available on the Internet at http://www.cerncourier.com/main/article/39/10/8/1.

SLIDE #17

Evidence That Relativistic Halo Protons Could Be the Long-Sought Dark Matter

- The required gravitational strength for dark matter could be achieved even with **too few** protons determined "according to the calculations of particle synthesis during the Big Bang" [Cline] because the very high relativistic mass of the cosmic-ray protons will amplify their gravitational attraction. [Drexler]

- Inflation theory of cosmologists requires a mass density in the Universe much greater than rest-mass protons could provide, but that relativistic-mass protons can provide. [Drexler]

- Protons play a more key role than helium nuclei since there are ten times as many protons as helium nuclei, and UHE protons are more relativistic since they move twice as fast as helium nuclei at the same kinetic energy. Thus, although the rest mass of a proton is one-fourth that of helium nuclei, a UHE proton's relativistic mass can be orders of magnitude greater than that of helium nuclei. For this reason, relativistic-mass abundances could be very different from rest-mass abundances.

SLIDE #18

Evidence That Relativistic Halo Protons Could Be the Long-Sought Dark Matter

- The highest energy cosmic rays are protons. Helium nuclei, electrons, and a few heavier nuclei are also moving through the galaxy as cosmic rays.

- About 1,000 cosmic ray relativistic protons strike each square meter of the Earth's atmosphere per second.

- Primary proton cosmic rays span a large range of kinetic energies from 10^{10} electron volts to at least 10^{20} electron volts.

SLIDE #19

Evidence That Relativistic Halo Protons Could Be the Long-Sought Dark Matter

- Only three or four primary cosmic rays per square kilometer per century reach energies above 10^{19} electron volts (approximately the kinetic energy of a fast-pitched baseball).

- Thus far, there are no clear directional clues from primary cosmic rays indicating the direction of their source.

- Measurements indicate that the highest energy primary cosmic rays are protons that exhibit energies above 10^{19} electron volts.

- The kinetic energies of many primary cosmic rays are many orders of magnitude higher than the 10^{12} electron volt particles produced by man-made accelerators.

SLIDE #20

Additional Evidence of UHE Proton Dark Matter: Relativistic Halo Protons Create Beryllium and Boron

- Scientists believe that ultra-high-energy relativistic protons exist in the halo surrounding the Milky Way (a spiral galaxy). These protons are credited with creating beryllium and boron by smashing into heavier elements, tearing them apart and thereby creating those two lighter elements.

- Since beryllium and boron are found in the low metallicity halo stars, that provides evidence that UHE relativistic protons do exist in the halos of spiral galaxies. [Drexler]

SLIDE #21

Additional Evidence of UHE Proton Dark Matter: The Source of Cosmic-Ray Protons Surrounds the Earth

- NASA's website states, "If you made a map of the sky with cosmic ray intensities, it would be completely uniform."

- The University of Birmingham U.K. website states, "If cosmic rays come from outside of the Galaxy the spatial position of cosmic ray intensity would change at different times during the day, due to the fact that the Earth is spinning on its own axis. No such variation is observed. Indeed the cosmic ray intensity on all points on the Earth's surface is roughly the same, at all times during the day. Therefore it is thought cosmic rays have no spatial or time variation."

- A cosmic-ray source with the characteristics indicated above, in close proximity to the Milky Way must completely surround the solar system and the Earth. [Drexler]

SLIDE #22

There Is a Similar Ratio of Protons to Helium Nuclei in Cosmic Rays and in the Solar System and in the Universe

- Website information from the Laboratory for High Energy Astrophysics at NASA/Goddard Space Flight Center indicates that protons represent about 90% of the cosmic rays, and helium nuclei, about 9%.

- This ratio of 10:1 is close to the 12:1 anticipated from the Big Bang theory for the entire Universe.

- This provides a clue that the source of cosmic rays raining on the Milky Way is probably similar in nature to the source of cosmic rays raining on other galaxies.

SLIDE #23

PART III

THE DARK MATTER HALO OF SPIRAL GALAXIES

SLIDE #24

Linking UHE Protons and Cosmic-Ray Protons to Dark Matter Particles

- There may be some distinguishing features regarding the UHE protons passing through the border region between a galaxy and its dark matter halo that may disclose the nature of dark matter. [Drexler]

- The majority of cosmic-ray protons approaching the Earth have kinetic energies less than 10^{15} eV according to the Center for European Nuclear Research *CERN Courier.*

- Protons with energies of 10^{20} eV or higher are very rare.

- Like dark matter, the positively charged UHE cosmic-ray protons and helium nuclei are not detectable directly by any known type of telescope.

SLIDE #25

Linking UHE Protons and Cosmic-Ray Protons to Dark Matter Particles

- In 1984, A. M. Hillas* and in 1985, A. A. Watson* published data regarding the flux of cosmic ray protons on the Earth's atmosphere with energies ranging from 10^{11} eV to 10^{20} eV. Their data, confirmed by others, shows a "bump" in the energy spectrum of cosmic rays at about 10^{14} to 10^{15} eV. Then the flux falls dramatically for kinetic energies above about 5×10^{15} eV, which is called the "knee" in the cosmic-ray energy spectrum.

- This "knee" region is considered so important by astrophysicists that the ACCESS (Advanced Cosmic-Ray Composition Experiment for the Space Station) experiment was designed to investigate the "knee" of the energy spectrum curve by direct measurement of cosmic rays. Part of the ACCESS concept is a transition radiation detector that was scheduled for prototype test at CERN in late 2001.

- A possible explanation for the "knee" in the energy spectrum is in slides #28 and #30.

*A. M. Hillas (1984) Ann. Rev. Astr. Astrophys, 22, 425 and A. A. Watson (1985) 19th Intl. cosmic ray conference, La Jolla, USA, vol. 9, page 111.

SLIDE #26

Linking UHE Protons and Cosmic-Ray Protons to Dark Matter Particles

Source: *Cosmic Bullets – High Energy Particles in Astrophysics*, by Roger Clay and Bruce Dawson (1997), pp. 87-88, "Properties of Primary Cosmic Rays":

- "The [particle] cosmic rays are deflected less [by magnetic fields in the Galaxy] as their [kinetic] energy increases and they follow great looping spiral paths, the scale of which increases in proportion to the cosmic ray [particle kinetic] energy...."

- "This scale begins to be comparable with the size of our Galaxy at cosmic ray energies of over 10^{15} eV. Below such energies we can be reasonably confident that the particles wander more or less randomly within the Milky Way. We tend to think of these as being trapped within the Galaxy...."

- "At higher energies [than 10^{15} eV], we assume that the cosmic rays leave our Galaxy quite quickly..." stated Clay and Dawson.

SLIDE #27

Linking UHE Protons and Cosmic-Ray Protons to Dark Matter Particles

- Those protons with 10^{17} eV > E > 10^{15} eV might leave the galaxy quickly, but they probably would be trapped in the dark matter halo which extends 10 to 20 times beyond the size of the galaxy.* [Drexler]

- The relativistic masses of those UHE halo-trapped protons would range between $1,100,000 m_o$ to $110,000,000 m_o$ for protons with kinetic energies at 10^{15} eV and 10^{17} eV. Such trapped massive protons could create the powerful gravitational field in the halo normally associated with dark matter. [Drexler]

- Since most protons with 10^{17} eV > E > 10^{15} eV probably would be trapped in the dark matter halo, the cosmic-ray protons arriving at the Earth would be almost entirely E < 10^{15} eV galaxy protons. That might explain the "knee" in the energy spectrum (slides #17 and #30) and the rapid decline with kinetic energy of the flux of cosmic-ray protons (slides #16 and #17). [Drexler]

*The entrapment of protons in the halo or in the galaxy, for some protons, may last for millions of years and for others perhaps billions of years, after which they may become cosmic-ray protons. [Drexler]

SLIDE #28

February 13, 2003 *Nature* Article Reports Discovery of 10^{12} Solar Masses of Hot "Missing" Intergalactic Baryons That Stabilize the Local Group of Galaxies

New evidence suggesting proton/baryonic dark matter:

- "The far-ultraviolet signature of 'missing' baryons in the Local Group of galaxies," *Nature*, February 13, 2003.

- The missing baryons were detected by "UV absorption lines in the spectra of background sources such as quasars."

- "Up to two-thirds of 'missing' baryons may have escaped detection because of their high temperature and low density." *

- "At least 82% of these [UV] absorbers [baryons] are not associated with any 'high velocity' atomic hydrogen complex in our Galaxy, and are therefore likely to result from a primordial warm-hot intergalactic medium pervading an extended corona around the Milky Way or the Local Group [of galaxies]."

*The temperature of this low density intergalactic "fog" ranged between 100,000 to 10 million degrees Kelvin.

SLIDE #29

Astronomical Evidence Supports UHE Protons as a Principal Component of Dark Matter [Drexler]

- The linking of the "knee" in the cosmic-ray energy spectrum of slides #17 and #28 to the rapid decline with kinetic energy of the flux of cosmic-ray protons in slides #16 and #17 and the concept of UHE proton trapping in the dark matter halo and in the galaxy in slides #27 and #28 seems to create a self-consistent astronomical- and astrophysical-based cosmology supporting UHE protons as a dark matter candidate.

- *Nature* (February 13, 2003) reports discovery of 10^{12} solar masses of hot "missing" intergalactic baryons that pervade the Milky Way and the Local Group of galaxies, which stabilizes the Local Group galaxies. According to the article, it is "not associated with any 'high velocity' atomic hydrogen complex in our Galaxy." This new data provides evidence of hot UHE proton dark matter. (See slide #29.)

- Beryllium and boron in low-metallicity halo stars provide strong evidence of massive UHE halo protons, which scientists believe smashed heavier elements to create the two lighter elements. (See slide #21.)

SLIDE #30

Particle Physics Evidence Supports UHE Protons as a Principal Component Of Dark Matter [Drexler]

- The relativistic masses of UHE halo-trapped protons would range between 1,100,000m_o to 110,000,000m_o for protons at 10^{15} eV and 10^{17} eV. Such trapped massive protons could create the powerful gravitational field in the halo normally associated with dark matter. (Protons with E < 10^{15} eV are trapped in the galaxy.)

- The very-high and ultra-high energies of cosmic-ray protons reaching the Earth's atmosphere provide evidence of sources of UHE protons located in both the galaxy and the halo. Apparently, a small percentage of these UHE-trapped protons escape as cosmic-ray protons on a regular basis.

SLIDE #31

PART IV

ARE UHE DARK MATTER HALO PROTONS RELICS OF THE BIG BANG?

SLIDE #32

Are UHE Dark Matter Protons Relics of The Big Bang? There Appears to Be Support for That Theory

It is generally believed dark matter, the source of dark matter, or the dark matter components were created by the Big Bang. To strengthen the theory of UHE dark matter protons, it would be desirable to find a credible method for creating UHE protons with energies of at least 10^{20} eV by the Big Bang mechanism.

References follow:

- Alan H. Guth in *The Inflationary Universe* said, "To make contact with the grand unified theories that Henry Tye was trying to persuade me to work on, one would have to trust the extrapolation all the way back to 10^{-39} seconds after the Big Bang when the temperature was 10^{29} °K. At that temperature the average energy per particle would be about 10^{16} GeV" [or 10^{25} electron volts].

- From Kings University College, Alberta Canada – At 10^{-38} seconds after the Big Bang the temperature reached 10^{29} °K, making the average energy per particle 10^{25} electron volts.

SLIDE #33

Are UHE Dark Matter Protons Relics of The Big Bang? There Appears to Be Support for That Theory

- *Europhysics News* (2002), Vol. 33, No. 5 reports "The mass of particles should be typically in the Grand Unification Theory energies ($\sim 10^{25}$ eV). They would have been created when the temperature of the Universe was of the same order of magnitude, i.e. roughly 10^{-35} second after the Big Bang. They would have survived up to now by some yet unknown mechanism (a very weakly violated quantum number, particles trapped inside huge potential wells called topological defects...). They would have accumulated by gravitational attraction in the halo of galaxies (therefore escaping the GZK cutoff)." *

- The 1966 GZK theory was widely accepted until 1998, when Sidney Coleman and Sheldon L. Glashow published a paper entitled, "Evading the GZK cosmic-ray cutoff." They point out that "Tiny departures from Lorentz invariance, too small to have been detected otherwise, have effects that increase rapidly with energy and can kinematically prevent cosmic-ray nucleons from undergoing inelastic collisions with CBR photons. If the cut-off is thereby undone, a deeply cosmological origin of UHE cosmic rays would become tenable."

*See footnote on slide #35.

SLIDE #34

Are UHE Dark Matter Protons Relics of The Big Bang? There Appears to Be Support for That Theory

- The *Europhysics News* (2002, Vol. 33, No. 5) article (see slide #34) supports the UHE dark matter proton theory by indicating that high energy particles created by the Big Bang "would have survived up to now...."

- Coleman and Glashow also provide support for the Big Bang origin of UHE cosmic rays by stating, "If the GZK cutoff is thereby undone, a deep cosmological origin of UHE cosmic rays would become tenable" and that cosmic-ray energies have been observed well beyond 5×10^{19} eV.

*According to the currently questioned 1966 Greisen-Zatsepin-Kuzmin (GZK) cutoff theory, protons with energies greater than 6×10^{19} eV would interact with the cosmic microwave background radiation and lose energy through radiation and thus would not travel more than 50Mpc, or about 160 million light-years (pc = parsec = 3.26 light-years). See slide #34 for an update on the GZK theory.

<u>SLIDE #35</u>

It Has Been Assumed, Without Proof, That *All* Big Bang UHE Protons Became Hydrogen Atoms 700,000 Years Later

- Astrophysics Research Program of U.C. Berkeley – "...it took about 700,000 years of cooling until this was able to occur. The capture of electrons to form atoms resulted in an important change in the universe. At that moment, without free electrons to interact with the photons present, the universe became transparent to radiation."

- Department of Physics and Astronomy, University of Tennessee – "As the free electrons are bound up in atoms the primary cross section leading to the scattering of photons (interaction with the free electrons) is removed and the Universe (which has been very opaque until this point) becomes transparent...."

- A. Kosowsky of Cal Tech – "As the universe drops below a temperature of around 0.1 eV at a red shift of around 1300, the electrons and protons begin to 'recombine' into neutral hydrogen. Within a short time, almost all the free electrons are converted to neutral hydrogen...."

- Today's multitudinous cosmic-ray protons and UHE protons provide contrary evidence. [Drexler]

SLIDE #36

The 10^{25} eV UHE Protons Created by the Big Bang Declined in Energy by Seven to Ten Orders of Magnitude Over 13.7 Billion Years

- In the halo of the Milky Way the UHE proton energies are estimated to range between 10^{15} eV and 10^{17} eV. Since immediately after the Big Bang their average energies are estimated to be in the range of about 10^{25} eV, the energy decline over the past 13.7 billion years appears to be at least 100 million and as much as 10 billion. The reasons for this enormous energy (and mass) decline are discussed in slides #39, #40, #42, and #43.

- The 10^{15} eV and 10^{17} eV protons have been confined to the galaxy's dark matter halo by the extragalactic magnetic field (and gravitational attraction) to a much greater extent than uncharged hydrogen atoms would be confined under gravitational attraction alone. (See slide #85.)

SLIDE #37

PART V

THE UHE PROTONS IN THE HALOS OF SPIRAL GALAXIES STILL RETAIN ENORMOUS ENERGIES

SLIDE #38

The 10^{25} eV Big Bang UHE Protons Have Fallen in Energy by a Factor of 100 Million to 10 Billion Over 13.7 Billion Years; Probable Cause: Synchrotron-Radiation Energy Loss

- Synchrotron radiation is electromagnetic radiation that is emitted by charged particles moving at relativistic velocities in circular or spiral orbits in magnetic fields. The rate of radiation emission is inversely proportional to the product of the radius of curvature of the particles' orbit and the fourth power of the mass of the particles. Therefore, synchrotron radiation from a proton is only 8.78×10^{-8} of that from an electron. For UHE protons, the radiation can be in the form of X rays or gamma rays.

- The radius of curvature of the path of a proton crossing orthogonal magnetic flux lines is equivalent to the Larmor radius (see slide #85) which is directly proportional to the kinetic energy of the proton and inversely proportional to the strength of the magnetic field. Thus, the synchrotron radiation from a UHE proton is inversely proportional to the proton's kinetic energy and directly proportional to the orthogonal magnetic field it traverses.

SLIDE #39

The 10^{25} eV Big Bang UHE Protons Have Fallen in Energy by a Factor of 100 Million to 10 Billion Over 13.7 Billion Years; Probable Cause: Synchrotron-Radiation Energy Loss

According to James W. Cronin* the extragalactic magnetic field is $\leq 1 \times 10^{-9}$ gauss and the magnetic field of the Milky Way is 2×10^{-6} gauss, or 2,000 times greater. Thus, as indicated in slide #39, a UHE proton moving within the Milky Way would generate synchrotron radiation at a rate 2,000 times greater than when it is traveling in the extragalactic field, and therefore its kinetic energy will decline accordingly. As the proton's kinetic energy falls, the synchrotron radiation would further increase, accelerating the proton's energy losses. Such a decelerating proton should eventually arrive at a star system, such as our solar system in the Milky Way, as a cosmic-ray proton.

*Unsolved Problems in Astrophysics, John N. Bahcall and Jeremiah P. Ostriker, Editors, Princeton University Press, 1997, page 332.

SLIDE #40

PART VI

UHE PROTONS CREATE SYNCHROTRON RADIATION IN THE FORM OF GAMMA RAYS

SLIDE #41

Milky Way Galaxy With Gamma-Ray Halo – New Class of Gamma Ray Objects Discovered in Milky Way

NASA's March 22, 2000 news release is entitled, "New Class of Gamma Ray Objects Discovered In Milky Way." Key points are as follows:

- "The exotic world of gamma-ray astronomy has taken yet another surprising turn with the revelation that half of the previously unidentified gamma-ray sources in our own galaxy, the Milky Way, actually comprise a new class of mysterious objects."

- "These are objects we've never seen before," said Dr. Neil Gehrels, an astrophysicist at NASA Goddard Space Flight Center and lead author on the *Nature* [March 23, 2000] article. "We can't make out what they are yet, but we know they're different and, boy, there's a lot of them."

- Gehrel's co-authors of the *Nature* article are Drs. Daryl Macomb, David Bertsch, David Thompson, and Robert Hartman, all of NASA Goddard. The article said the new class of objects is very different from the famous gamma-ray burst sources **because the gamma rays shine continuously instead of coming in a flash**, like the normal gamma-ray bursts.

SLIDE #42

"Gamma-Ray Glow Bathes Milky Way" – Gamma-Ray Emission From UHE Cosmic Rays

- Gamma rays are the most powerful form of radiation. They are generated in several ways: Gas at a temperature of 10 billion degrees glows in gamma rays; also, **energetic particles smashing into other particles or spiraling through magnetic fields release gamma rays.**

- *Science News*, November 8, 1997, page 292, entitled, "Gamma-ray glow bathes Milky Way," states: "A mysterious halo of gamma rays not associated with any known celestial objects extends thousands of light-years from the core of the Milky Way and may surround the entire galaxy, astronomers report."

- "These gamma rays are providing the first evidence that some sort of high-energy process is occurring at large distances from the [Milky Way] galactic core," said physicist David D. Dixon of the University of California, Riverside. "The gamma-ray distribution may also provide indirect evidence of dark matter – the universe's missing mass, whose existence scientists have inferred but not yet demonstrated," Dixon said.

<u>SLIDE #43</u>

"Gamma-Ray Glow Bathes Milky Way" – Gamma-Ray Emission From UHE Cosmic Rays

- T. Totani's article in *Astron. Astrophys. Suppl. Ser.* 142, 443-445 (January 3, 2000) is entitled, "TeV gamma-ray emission from gamma-ray bursts and ultra high energy cosmic rays." ... "We show that such extreme phenomena can be reasonably explained by synchrotron radiation of protons accelerated to ~10^{20-21} eV. There may also be TeV emission in afterglow phase from external shocks, and proton synchrotron in this phase gives a quantitative explanation for the famous long duration GeV emission from GRB 940217."

- Totani continued, "Both experiments suggest a time scale of 10 TeV emission as ~10 seconds. Considering the fact that two different experimental groups independently reported similar significant signals of 10-20 TeV gamma-rays, these observations are now worth theoretical interpretations. The famous **long duration GeV emission** [emphasis added] from GRB 940217 (Hurley et al. 1994) **can be explained by the proton synchrotron radiation in the GeV range** [emphasis added] in the early afterglow phase (Totani 1998a)."

SLIDE #44

X Ray Synchrotron Radiation From UHE Electron Cosmic Rays – ASCA Report Entitled, "SN 1006, A Source Of Cosmic Rays"

- ASCA (Advanced Satellite for Cosmology and Astrophysics) was on an X ray mission 1993-2001. Synchrotron radiation from electron cosmic rays was detected and analyzed.

- SN1006 may be the first example of synchrotron radiation from electron cosmic rays with energies about 100 trillion electron volts (100 TeV) within a SNR's shell.

- How do we know it is synchrotron radiation? The spectrum from thermal X rays generally shows a characteristic set of lines, while **synchrotron radiation forms a hard continuum that has no emission lines.**

- To summarize, note from slides #42, #43, #44, and this slide that synchrotron radiation from either UHE protons or electrons forms a hard continuum that has no emission lines.

SLIDE #45

PART VII

BIG BANG ORIGIN OF 10^{20} eV COSMIC-RAY PROTONS FOUND IN THE MILKY WAY?

SLIDE #46

Big Bang Origin Is a Possible Explanation for 10^{20} eV Cosmic-Ray Protons

- James W. Cronin, in *Unsolved Problems in Astrophysics*, stated in 1997, "Recently, two unusual cosmic rays with energies more than 2×10^{20} eV have been reported. These events are extraordinary, there is no credible model for their acceleration. ... If, as is likely, these cosmic rays are extragalactic protons...." "The existence of these high-energy rays is a puzzle, the solution of which will be the discovery of new fundamental physics or astrophysics."

- Herbert Friedman, in *The Astronomer's Universe*, stated in 1998, "There is no known mechanism for accelerating cosmic rays to 10^{20} electron volts...."

- M. S. Longair, in *High Energy Astrophysics*, volume 2, chapter 20.5, "The Origin of Cosmic Rays," stated in 1994, "It is, therefore, quite feasible that the highest energy cosmic rays are indeed extra-galactic, and they could come from radio galaxies, alive or extinct within the local super-cluster.... In the present instance, the most straightforward interpretation of the data is that the bulk of the cosmic rays are of Galactic origin but that the very highest energy particles, with $E \geq 10^{19}$ eV, may have an origin outside our Galaxy and probably within the local supercluster."

<u>SLIDE #47</u>

Big Bang Origin Is a Possible Explanation For 10^{20} eV Cosmic-Ray Protons Found in the Milky Way

- A. V. Olinto (Enrico Fermi Institute, University of Chicago), in her March 1, 2000 paper entitled, "The Mystery of Ultra-High Energy Cosmic Rays," said, "The origin of cosmic rays with energies above 10^{20} eV is an intriguing mystery. At present, about 20 events above 10^{20} eV have been reported worldwide by experiments such as the High Resolution Fly's Eye, AGASA [Akeno Giant Air Shower Array]...."

- P. Bhattacharjee and N. Gupta (India Institute of Astrophysics), in their May 2003 paper entitled, "Ultrahigh Energy Cosmic Rays and Prompt TeV Gamma Rays from Gamma Ray Bursts" stated, "The origin of the observed Ultrahigh Energy Cosmic Ray (UHECR) events with estimated energy in excess of 10^{20} eV is unknown and is currently a subject of much discussions. It is generally extremely difficult to accelerate particles to such high energies in most of the known astrophysical objects through conventional acceleration mechanisms."

SLIDE #48

PART VIII

THE ACCELERATING EXPANSION OF THE UNIVERSE AND DARK ENERGY

In an expanding Universe, all galaxy clusters are moving away from each other at some velocities. When those separation velocities continue to increase, the Universe has entered into an accelerating expansion phase, which has been attributed to *"dark energy."*

SLIDE #49

The Accelerating Expansion of The Universe and Dark Energy

- In 1929, Edwin P. Hubble announced that with the exception of the galaxies closest to the Milky Way, the galaxies are rushing away from each other in all directions and, therefore, the Universe is expanding. His astronomical evidence led to the Big Bang theory of the creation of the Universe.

- In 1998, a ten-year study led by Dr. Saul Perlmutter of spectacular astronomical events involving exploding stars (supernovae) led to the discovery that the expansion of the Universe was accelerating. Those measurements were made by ground-based telescopes. Recently, Type 1a supernovae were studied with the Hubble space telescope that confirmed and refined the 1998 conclusions. Perlmutter, of the Lawrence Berkeley National Laboratory, is leader of the International Supernova Cosmology project.

- The term "*dark energy*" was selected by cosmologists as the name for the hypothetical form of energy that permeates all space and has negative pressure resulting in a repulsive gravitational force that accelerates the velocity of separation of galaxy clusters.

SLIDE #50

The Accelerating Expansion of The Universe and Dark Energy; May 30, 2003 *New York Times* Article

Following the May 2003 meeting of the American Astronomical Society, John Noble Wilford of *The New York Times* wrote a May 30, 2003 article entitled, "From Distant Galaxies, News of a Stop-and-Go Universe." Excerpts are reproduced in slides #51, #52, and #53:

- "Once the universe was expanding at a decelerating rate but then began accelerating within the last seven billion years."

- "The combined gravitational pull from all matter in the universe, most of which is beyond detection, has acted as a brake on cosmic expansion. The so-called dark matter apparently had the advantage when the universe was younger, smaller and denser. Now the ever-increasing pace of expansion suggests that something else even more mysterious is at work. Theorists are not sure what the antigravity force is, but they call it dark energy. It has apparently gained the upper hand."

SLIDE #51

The Accelerating Expansion of The Universe and Dark Energy
May 30, 2003 *New York Times* Article

- "For the current research, astronomers observe what are called Type 1a supernovae, stellar explosions that at their peak are brighter than a billion stars like the Sun. They are thus visible across billions of light-years of space, and a close examination of their light reveals the distance, motions and other evidence of conditions. As the light travels to Earth, the wavelengths are stretched by an amount that reflects the universe's expansion when the star exploded."

- "Dr. Robert P. Kirshner of the Harvard-Smithsonian Center said the four extremely distant supernovae indicated that the universe seven billion years ago was 'in fact winning this sort of cosmic tug-of-war,' but now dark energy is more dominant."

- "Scientists said they assumed that with the stretching out of space the proportion of dark energy to dark matter had been reversed. In the earlier and denser universe, matter of all kinds, the invisible dark matter and the visible ordinary matter of stars and planets, predominated."

SLIDE #52

The Accelerating Expansion of The Universe and Dark Energy

- *New York Times*, May 30, 2003: "The team of Dr. John Tonry and Dr. Robert Kirshner estimates that about 60 percent of the universe is filled with dark energy and 30 percent of the mass is dark matter. The remaining 10% consists of ordinary matter, only 1 percent of which is visible in the galaxies. Theorists offer roughly the same estimates and surmise that the changeover from dark matter to dark energy domination probably occurred before 6.3 billion years ago."

- On October 10, 2003 this estimate was changed to 5 billion years ago, by Dr. Adam Riess, an astronomer at the Space Telescope Science Institute in Baltimore, who described it as a "cosmic jerk." His report was presented at a meeting on the Future of Cosmology at the Case Western Reserve University in Cleveland.

SLIDE #53

The Accelerating Expansion of The Universe and Dark Energy

- In 2001, Michael S. Turner at the University of Chicago published a paper entitled, "Dark Energy and the New Cosmology," in which he stated, "Dark energy by its very nature is diffuse and a low-energy phenomenon. It probably cannot be produced at accelerators; it isn't found in galaxies or even clusters of galaxies. The Universe itself is the natural lab – perhaps the only lab – in which to study it. The primary effect of dark energy on the Universe is on the expansion rate."

- In 1995, Trinh Xuan Thuan in a book entitled, *The Secret Melody*, wrote, "Between the age of 300,000 years and today, the distance between galaxies has increased by a factor of 1,000 and the average density of the universe has decreased by a factor of one billion." He also pointed out that when local gravity dominates galaxies, "they no longer followed the expansion [of the universe]...."

SLIDE #54

These Principal Points Used to Explain Dark Matter Also Will Be Used to Explain the Accelerating Expansion of the Universe

- The objective of the slide presentation to this point was to present a theory of dark-matter-halo massive particles based upon the known existence of UHE cosmic ray protons, each having a mass equivalent of thousands to hundreds of millions of protons at rest. (See slide #15.)

- One of the questions that required a plausible answer was, what is the origin of the UHE cosmic-ray protons. This is covered in slides #33, #34, #35, #47, and #48.

- The GZK cut-off question is answered by the 1998 and 2002 analyses of the 1966 GZK cut-off theory as presented in slides #34 and #35. The 1998 paper by Coleman and Glashow points out that inelastic collisions with CBR photons decline rapidly with the energy of cosmic-ray nucleons. The UHE proton certainly fits within that category. The 2002 *Europhysics News* report (see slides #34 and #35) indicates the Big Bang particles trapped in potential wells "would have survived up until now."

SLIDE #55

These Principal Points Used to Explain Dark Matter Also Will Be Used to Explain the Accelerating Expansion of the Universe

- Contrary to the opinions from three universities (slide #36), it is unlikely that essentially all of the Big Bang UHE protons have been converted to hydrogen atoms. The existence of multitudinous cosmic-ray protons and UHE protons provides contrary evidence. (See slides #16, #17, #21, and #26 – #30.) [Drexler]

- The estimate of the Big Bang UHE protons losing an extremely large percentage of their original kinetic energy to synchrotron radiation is discussed in slides #39 and #40.

- If UHE protons do represent the dark matter massive particles in the halo of spiral galaxies, their synchrotron radiation should be detectable as long duration emission of gamma rays in the halo of the Milky Way. In recent years, they have been observed. Long-duration gamma-ray emission in the form of a halo around the Milky Way was reported by scientists in the 1997-2000 time period as mentioned in slides #42, #43, and #44.

SLIDE #56

These Principal Points Used to Explain Dark Matter Also Will Be Used to Explain the Accelerating Expansion of the Universe

- In order for UHE halo protons to have typical energies of 10^{15} eV to 10^{17} eV today, they would have had to have very much higher energies at the time of the Big Bang. Slides #33, #34, and #35 provide literature references that Big Bang protons were created with an average energy of 10^{25} eV.

- The origin of proton cosmic rays with energies today near 10^{20} eV are considered a mystery or a puzzle by the world's leading astrophysicists for various reasons including those discussed in slides #33, #34, #35, #37, #39, and #40. Most researchers and authors believe that the origin of 10^{20} eV cosmic ray protons is extragalactic as evidenced by their comments in slides #47 and #48.

SLIDE #57

The Accelerating Expansion Between Galaxy Clusters (Not Between Galaxies)

- Note that Michael Turner points out (see slide #54) that the dark energy "isn't found in galaxies or even clusters of galaxies." This is an extremely important point. This statement means that dark energy pushes galaxy clusters apart but doesn't push galaxies apart within clusters and doesn't push stars apart within a galaxy.

- This is observed locally. While the Universe is expanding, in our Local Group of galaxies Milky Way and Andromeda are moving towards each other at 119 km/sec. Also, our Local Group is moving toward the Local Supercluster at 600km/sec. This illustrates that local strong gravitational forces can overcome the dark energy antigravity forces. This also suggests that perhaps in the earlier, smaller, and denser Universe, when the Universe was less than about 8.2 billion years old, the closeness of the galaxy clusters favored dark matter gravity over dark energy antigravity, and expansion acceleration could not take hold. When the Universe was 8.7 billion years old, the expansion acceleration began, according to Adam Riess. (See slide #53.)

SLIDE #58

Jerome Drexler's Theory of the Accelerating Expansion Between Galaxy Clusters

- As the synchrotron radiation emission of gamma rays continues, not only does the kinetic energy of the halo UHE protons fall, but their relativistic mass will fall as well (slides #39 and #40). See slide #15 and assume that the average kinetic energy of the UHE protons declined from 10^{16} eV to 5×10^{15} eV over a period of time in the dark matter halo of some galaxy cluster. This represents a decline in the dark matter mass in the halo of that galaxy cluster of 50% and perhaps a 40% decline in the mass of the combined galaxy cluster and its halo.

- Such a reduction in the dark matter halo mass around galaxy clusters would:

    (1) Raise the galaxy clusters' velocities under the Law of Conservation of Linear Momentum. (See footnote on slide #75.) [Drexler]

    and

    (2) Reduce each galaxy cluster's gravitational attraction to nearby galaxy clusters, thereby facilitating their more rapid separation. [Drexler]

SLIDE #59

Drexler's Theory of the Accelerating Expansion Between Galaxy Clusters

- The gravitational attraction effect of item (2) on the previous slide will diminish through the years as the nearby galaxy clusters become more distant. Note from slide #54 that from the age of 300,000 years until today, the spacings between galaxies increased by a factor of 1,000.

- In an expanding Universe, all galaxy clusters are moving away from each other. Meanwhile, the masses of their dark matter halos of UHE protons are declining because of the synchrotron radiation energy losses. As a result, the velocity of every galaxy cluster should rise (owing to the reduced gravitational attraction between them and the Law of Conservation of Linear Momentum), thereby accelerating the expansion of the Universe. [Drexler]

SLIDE #60

Drexler's Theory of the Accelerating Expansion Between Galaxy Clusters

- The antigravity repulsion between galaxy clusters is proportional to the relative decrease in mass of their dark matter halos of UHE protons owing to the protons' loss of kinetic energy through synchrotron radiation. [Drexler]

- Gravitational attraction is directly proportional to the product of the masses of two galaxy clusters and inversely proportional to the square of the distance between them.

- Thus, for the earlier, smaller, and denser Universe more than five billion years ago, the smaller distances between galaxy clusters caused the gravitational attraction between them to be high because of the inverse square of those smaller distances between galaxy clusters. This inverse-square relationship may be a principal reason that the accelerated expansion did not begin until five billion years ago when the conservation-of-momentum effect and reduced galaxy cluster mass finally overcame the gravitational attraction between galaxy clusters. [Drexler]

SLIDE #61

PART IX

JEROME DREXLER'S THEORY OF STAR AND SUN FORMATION

UHE cosmic-ray protons and cosmic-ray helium nuclei may have helped create the mass and fuel of the Sun and trigger its fusion reaction.

SLIDE #62

Drexler's Theory:
What Is the Source of About 150 Billion Solar Masses of Hydrogen That Created The Stars of the Milky Way?

- The Milky Way contains an estimated 400 billion stars (plus or minus 200 billion). Those stars represent a mass of about 175 billion suns, and 96% of them are thin-disk stars located close to the galactic plane. Where is the source of those 175 billion solar masses of hydrogen?

- Jan Oort has estimated that as much as one solar mass of hydrogen (containing helium and dust) flows out of the Milky Way's central bulge each year into the spiral arms of the galaxy's thin extended disc where the Sun is located.

- If that same hydrogen flow into the spiral arms occurred over a 13-billion-year period, that would total only about 13 billion solar masses of hydrogen compared with about 150 billion solar masses of hydrogen in the stars in the Milky Way's thin disc that were born during that 13-billion-year period.

SLIDE #63

Drexler's Theory:
What Is the Source of About 150 Billion Solar Masses of Hydrogen That Created The Stars of the Milky Way?

- To arrive at that answer, consider the formation of one of those 400 billion stars, namely the Sun, which is a star in the thin disk of the Milky Way along with Sirius, Vega, Rigel, Betelgeuse, and Alpha Centauri. Ages of the thin-disk stars range between newborns to 10 billion years old and are known as Population I stars.

- The ultra-high-energy relativistic protons (hydrogen nuclei), helium nuclei, and heavier nuclei called cosmic rays that arrive at the planet Earth each day may have been showering our region of the Universe for most of the Universe's 13.7 billion years of existence.

SLIDE #64

Drexler's Theory:
Herbert Friedman said, "All Told, About 1-2 Pounds Per Day [of Cosmic-Ray Nuclei] Arrive at the Earth"

- Herbert Friedman, in his 1998 edition of *The Astronomer's Universe*, and Michigan State University (in 1993) have reported that a total of about 1 to 2 pounds of high-energy, cosmic ray protons, helium nuclei, and a few heavier nuclei arrive at Earth each day.

- If their origin is the dark matter halo around the Milky Way, that same flux of relativistic protons and helium nuclei would be expected over the entire solar system.

- Is it possible that the long-time showering of those high-energy protons, helium nuclei, and heavier nuclei onto the region now known as the solar system, along with a "dirty mix" of dust, heavy elements, and hydrogen-helium gas, would have been adequate to create the mass of the Sun, Jupiter, Saturn, and Earth? I utilized simple mathematics to explore that possibility.

SLIDE #65

Drexler's Theory: The Sun Came Into Being About 9 Billion Years After the Big Bang

- The Sun is 4.5 billion years old. (Its magnetic field is 50 gauss.) Since the age of the Universe is 13.7 billion years, there would have been available 9 billion years after the Big Bang for the Sun to accrete the cosmic ray protons and helium nuclei showering upon the region of the galaxy now known as the solar system. The outer reaches of the solar system are sometimes considered to be the Oort Cloud where billions to trillions of comets are held in orbit by the Sun's gravitational field.

- The Sun's gravitational field may also keep cosmic-ray particles within the solar system long enough for the Sun's magnetic field to spiral the cosmic-ray protons into a "death spiral" (see slides #79 and #80) followed by gravitational capture by the Sun. The Oort Cloud region is estimated to be located more than 50,000 A.U. (astronomical units) from the Sun, or at least 7.48×10^{12} kilometers, so that distance would represent the radius of a circle that encompasses the outer reaches of the solar system.

SLIDE #66

Drexler's Theory:
The 1 to 2 Pounds Per Day of UHE Protons and Other Nuclei Striking the Earth Means About 2×10^{18} Pounds Per Day Striking the Solar System

- Assume that during those initial 9 billion years, the flux of the cosmic-ray nuclei was equal to the 1 to 2 pounds per day currently striking the Earth's surface, which adds up to 166 to 332 kilograms per year. The Earth has a radius of 6.38×10^3 kilometers.

- The ratio of the circular area of the entire solar system (extended to the Oort Cloud) to the circular cross-sectional area of the Earth is 1.37×10^{18}, so the mass of UHE nuclei showering the extended solar system area each year would range between 2.27×10^{20} kilograms to 4.54×10^{20} kilograms.

- To determine the total mass of UHE protons and helium nuclei accreted during the initial 9 billion years, multiply the annual accretion by 9 billion years to arrive at an estimate of 2×10^{30} kilograms to 4×10^{30} kilograms as of 4.5 billion years ago, when the Sun was born. (The mass of UHE nuclei is determined from the Theory of Relativity by dividing their energies by c^2 as was done in slide #15.)

SLIDE #67

Drexler's Theory:
Maybe the UHE Cosmic-Ray Nuclei Did Create the Sun

- By adding the mass accretion from the time of the birth of the Sun until the present, the mass possibly residing within the solar system after 13.5 billion years is estimated to total between 3×10^{30} to 6×10^{30} kilograms. (Contribution to the Sun from the Milky Way bulge into the spiral arms of the galaxy would have added several percent of a solar mass, including iron which is necessary for life.) (Slide #70 points out the Sun has lost 20% of its mass due to the solar wind.)

- It is interesting to note that the actual mass of the Sun today as measured by astronomers is 2×10^{30} kilograms compared to the 3×10^{30} to 6×10^{30} kilograms estimated from accretion of UHE cosmic-ray nuclei onto the area of the extended solar system. Not only are the two mass quantities close, but the Sun's mass is lower, which is a more plausible expectation. [Drexler]

SLIDE #68

Drexler's Theory: From the Sun to Population I Stars In Other Spiral Galaxies

- Did the cosmic-ray protons, helium, and heavier particle nuclei actually create the Sun and the solar system? If they didn't, where is that 3×10^{30} to 6×10^{30} kilograms of protons, helium, and heavier nuclei hiding in the solar system region?

- If cosmic-ray protons did create the Sun, then their source, the halo UHE protons, should be a strong candidate for the long-sought dark matter particles.

- If they did create the Sun, cosmic-ray protons, helium, and heavier particle nuclei probably created the other thin-disc Population I stars in the galaxy as well. For Population I stars, the gestation/accretion periods would be estimated at more than 3 billion years, since the Population I stars in the galaxy are no more than 10 billion years old and the age of the Universe is about 13.7 billion years.

- It would then follow logically that Population I stars in the thin disc of other spiral galaxies are probably formed in a similar manner.

SLIDE #69

Drexler's Theory: The Sun's Mass May Be Greater Today Than at the Sun's Birth 4.5 Billion Years Ago

- According to the U.S. Department of Energy (DOE) the Sun converts 4×10^9 kilograms of mass into energy per second, which is only 5.7×10^{26} kilograms during the 4.5-billion year life of the Sun.

- However, the DOE also claims that the solar wind causes mass losses at a much higher rate, which may have resulted in a loss of 20% of the solar mass since its birth, or about 0.5×10^{30} kilograms.

- Compared to that weight loss, I estimated earlier that 1 to 2×10^{30} kilograms of nuclei were added to the solar system since the Sun's birth.

- Thus, the solar mass could be greater today than it was at the Sun's birth. If true, the masses of other Milky stars may be increasing. (These rough calculations assume that the current UHE cosmic ray nuclei flux of 1 to 2 pounds per day onto the Earth was constant over the entire solar system for the past 13+ billion years. The mass of the planets and other solar system objects totals only about 1% of the mass of the Sun.)

SLIDE #70

Drexler's Theory:
UHE Cosmic-Ray Nuclei May Have Facilitated the Triggering of the Sun's Fusion Reaction

- With the UHE cosmic-ray protons having a kinetic energy in the range of 10^{10} to 10^{20} electron volts, they not only added mass and fuel to the formation of the Sun but considerable nucleus-to-nucleus collision energy as well.

- These UHE nuclei provide a clue that the UHE protons and heavier high-collision-cross-section UHE nuclei may have facilitated the triggering of the Sun's fusion reactions and its birth.

- The traditional theory of the Sun formation involving a hydrogen gas cloud, forces of gravity, compression, and high temperature heating may have to be modified. (For further information about the UHE cosmic-ray nuclei, see slide #17, an energy distribution graph that is a plot of cosmic-ray particle flux versus particle energy.) [Drexler]

SLIDE #71

PART X

COSMIC-RAY COSMOLOGY:

DREXLER'S UNIFIED THEORY OF DARK MATTER, ACCELERATING EXPANSION, AND STAR FORMATION

Cosmic-Ray Cosmology Theory is based upon relativistic UHE protons moving through magnetic fields within the galaxy, the galaxy halo, and the Universe so as to create the dark matter gravitational strength, the accelerating expansion, and to participate in star formation.

SLIDE #72

Cosmic-Ray Cosmology: Drexler's Unified Theory of Dark Matter, Accelerating Expansion, and Star Formation

The Nature of Dark Matter

- UHE protons traveling near the speed of light have enormous relativistic mass and probably represent the dark matter mass in the halos of spiral galaxies and galaxy clusters. [Drexler]

- The mass needed to explain the enormous gravitational strength of the dark matter halo around spiral galaxies and galaxy clusters is derived from $m = E/C^2$ from Einstein's Theory of Relativity. That is, a proton traveling near the speed of light can have the gravitational attraction of one million or one billion stationary protons, as indicated in slide #15. [Drexler]

- The energy of the UHE protons probably is a relic from the 10^{25} eV protons believed to be created at the time of the Big Bang. [Drexler]

SLIDE #73

Cosmic-Ray Cosmology: Drexler's Unified Theory of Dark Matter, Accelerating Expansion, and Star Formation

Declining Dark Matter Mass Yields An Accelerating Universe

- The declining energy of the UHE protons since the Big Bang is attributed primarily to synchrotron radiation losses owing to the deflection (acceleration) of the high speed protons into spiral paths by the extragalactic and galactic magnetic fields. [Drexler]

- It is widely known that UHE protons deflected (accelerated) into circular or spiral paths by magnetic fields will lose kinetic energy through synchrotron radiation. From Einstein's Theory of Relativity, this means they are also losing mass; and, thus, the mass of the dark matter halos around spiral galaxies and galaxy clusters should be continually declining. [Drexler]

SLIDE #74

Cosmic-Ray Cosmology: Drexler's Unified Theory of Dark Matter, Accelerating Expansion, and Star Formation

Accelerating Expansion of the Universe

- It has been known for over 70 years that the Universe is expanding. Since 1998, it has been known that the expansion of the Universe is accelerating. That is, the galaxy clusters are moving away from each other at higher and higher velocities.

- If galaxy clusters are moving away from each other and if, in addition, the dark matter halo of each galaxy cluster is shedding mass isotropically, then the velocity of separation of the galaxy clusters should increase and the Universe expansion should accelerate owing to the Law of Conservation of Linear Momentum* and the reduced cluster-to-cluster gravitational attraction. [Drexler]

*The linear momentum of an object equals the product of its mass and velocity. If no external forces are acting on a group of galaxy clusters, the Law of Conservation of Linear Momentum requires that the total linear momentum of the galaxy clusters shall remain constant.

SLIDE #75

Cosmic-Ray Cosmology: Drexler's Unified Theory of Dark Matter, Accelerating Expansion, and Star Formation

The Nature of Star (Sun) Formation From Cosmic-Ray Protons

- *Star formation theory:* UHE protons, helium nuclei, and heavier nuclei probably created the Sun and the solar system. This is based upon the cumulative mass of cosmic-ray particles impinging upon and accreted by the extended solar system region over 9 billion to 13.5 billion years. Both the Sun's magnetic and gravitational fields probably played key roles in proton capture and the accretion process. (See slides #62 - #71 and #78 - #80.) [Drexler]

- *Protons to hydrogen:* It is unlikely that the Big Bang UHE protons were converted to hydrogen atoms 700,000 years later. The existence of multitudinous cosmic-ray protons and UHE protons, their apparent role in creating Population I stars, and the fact that dark matter is not comprised of hydrogen atoms, provide contrary evidence. [Drexler]

- *Cosmic-ray cosmology theory:* Slides #73 - #75 can be applied to dark matter, accelerating expansion, and star (sun) formation, without modifications, thus making it a unified cosmology theory. [Drexler]

SLIDE #76

PART XI

DREXLER'S THEORY OF "IMMORTAL" UHE PROTONS, "MORTAL" COSMIC-RAY PROTONS, AND THE "DEATH SPIRAL"

SLIDE #77

What Is the Difference Between a UHE Proton and a Cosmic-Ray Proton Bombarding the Earth?

- When a UHE proton in the halo of the Milky Way (a spiral galaxy) loses a significant portion of its kinetic energy over billions of years through synchrotron radiation, the proton will eventually plummet into the galaxy, thereby accelerating its energy loss. It then becomes one of the cosmic-ray protons which bombard wide regions of the galaxy, including the solar system, and may have played a role in creating the Sun and other stars as explained in slides #62 – #71.

- With UHE protons remaining active for billions of years, they may be thought of as "immortal" UHE protons while at end of their life they become "mortal" cosmic-ray protons plummeting into the galaxy in what I call a "death spiral."

SLIDE #78

The Transformation of "Immortal" UHE Protons Into "Mortal" Cosmic-Ray Protons Through the "Death Spiral"

- The synchrotron radiation loss of a relativistic charged particle is inversely proportional to both the radius of curvature of its path and the fourth power of its mass.

- The radius of curvature of a UHE proton's spiral path is equal to the Larmor Radius (see slide #85) and is directly proportional to the kinetic energy of the proton and inversely proportional to the magnetic field strength.

- The extragalactic magnetic field is reported to be about 1×10^{-9} gauss while the magnetic field in the interior of the Milky Way is about 2,000 times greater, at 2×10^{-6} gauss.

- Therefore, in the extragalactic dark matter halo of the galaxy, the magnetic field is very weak, the kinetic energy of the protons is high, the synchrotron radiation losses are extremely low, and the UHE proton may be able to circulate for billions of years. [Drexler]

SLIDE #79

The Transformation of "Immortal" UHE Protons Into "Mortal" Cosmic-Ray Protons Through the "Death Spiral"

- After billions of years in the extragalactic halo of a spiral galaxy, some of the UHE protons should eventually lose enough energy that their spiral paths will be reduced in diameter and the UHE protons will approach the surface of the galaxy. When the UHE halo protons enter the galaxy, their energy has perhaps diminished by a factor of 10 or so and the magnetic field might be about 100 times greater, thereby increasing the synchrotron radiation loss by a factor of 1,000. (The radius of a proton's spiral path is called the proton's Larmor Radius, which can be calculated as shown in slide #85.)

- Thus, as a UHE proton enters the galaxy, its energy will plummet further, say to half, thereby doubling the synchrotron radiation loss. The proton's energy will drop more and more, and it will enter into the "death spiral" as it plunges deeper into the galaxy and begins to be known as a cosmic-ray proton.

SLIDE #80

The Transformation of "Immortal" UHE Protons Into "Mortal" Cosmic-Ray Protons Through the "Death Spiral"

- The rate of synchrotron radiation is inversely proportional to the fourth power of the mass of the particle. Therefore, synchrotron radiation from protons is infinitesimal compared to synchrotron radiation from electrons. More precisely, proton synchrotron radiation losses are lower than radiation losses from electrons, following the same radius of curvature path, by a factor of about 11 trillion (the proton/ electron mass ratio of 1,836 to the fourth power).

- The discussion on slides #79 and #80 has to do with spiral galaxies, which are known to have a dark matter halo and also a black hole containing a few million solar masses.

SLIDE #81

PART XII

ASTRONOMERS REPORT ELLIPTICAL GALAXIES WITH NO DARK MATTER HALO – THEY GAVE NO REASONS

SLIDE #82

Astronomers Report Elliptical Galaxies With No Dark Matter Halo – They Gave No Reasons

- Elliptical galaxies are usually much larger than spiral galaxies and can have black holes containing about one billion solar masses or about 300 times the black hole mass of spiral galaxies.

- Cosmic-ray cosmology may provide an explanation for "naked" elliptical galaxies. If a galaxy has such a massive black hole, its high gravitational strength probably would shorten the active life of both UHE protons and cosmic-ray protons by pulling them into the elliptical galaxy and into the black hole relatively fast. Thus, any UHE proton dark matter that might have been present originally in a halo probably would have been consumed both by creating stars and by feeding the black hole. [Drexler]

SLIDE #83

PART XIII

DREXLER'S THEORY OF COSMIC-RAY COSMOLOGY APPLIED TO GALAXY FORMATION

SLIDE #84

Drexler's Cosmic-Ray Cosmology Applied to Galaxy Formation

The Proton Larmor Radius

- A proton crossing an orthogonal magnetic field enters into a spiral path. The radius of one cycle of that spiral path is called the Larmor Radius.

- The Larmor Radius of a proton can be calculated as:

$$r = 110 \text{ Kpc} \times \frac{10^{-8} \text{ gauss}}{B} \times \frac{E}{10^{18} \text{ eV}}$$

where Kpc means kilo parsec and one parsec equals 3.26 light-years

- The galactic magnetic field within the Milky Way is approximately 2×10^{-6} gauss compared to the extra-galactic magnetic field at 1×10^{-9} gauss.

Drexler's Cosmic-Ray Cosmology Applied to Galaxy Formation

The Proton Larmor Radius

- The Larmor Radius for a 10^{16} eV proton in the Milky Way halo's extragalactic magnetic field of 10^{-9} gauss is 11 Kpc; for a 10^{17} eV proton it is 110 Kpc; and for a 10^{18} eV proton it is 1,100 Kpc.

- The diameter of the Milky Way galaxy is about 100,000 light-years or 30.7 Kpc and its radius is about 15 Kpc.

- Studies in 1999 found that the dark matter halo of a spiral galaxy extends about 10-20 times the size of the visible regions (slide #3). Using a factor of 15, the radius of the dark matter halo would extend to perhaps 225 Kpc.

- Thus, some 10^{16} eV protons with a Larmor Radius of 11 Kpc might stay within the Milky Way galaxy's 15 Kpc radius. A 10^{17} eV proton with a Larmor Radius of 110 Kpc might remain within the halo's 225 Kpc radius, but a 10^{18} eV proton would probably leave both the galaxy and the halo.

SLIDE #86

Only One Dark Matter Candidate Establishes the Approximate Size Of the Milky Way

- From slides #85 and #86 it has been shown that the Larmor Radius for a 10^{16} eV proton in the Milky Way's extragalactic magnetic field of 10^{-9} gauss is 11 Kpc; for a 10^{17} eV proton it is 110 Kpc; and for a 10^{18} eV proton it is 1,100 Kpc.

- In slide #17 entitled, "Cosmic-Ray Energy Distribution for the Milky Way," the "knee" of the curve falls on a proton energy of approximately 5×10^{15} eV, which means that some protons with that energy might stay within the galaxy.

- From the above it could be concluded that the Milky Way galaxy should have a minimum radius of about 11 Kpc, compared to astronomers' estimate of about 15 Kpc.

- This seems to provide additional evidence that UHE protons are a credible dark matter candidate. No other currently proposed dark matter candidate can be used to estimate the size or even the order of magnitude of the size of the Milky Way galaxy.

SLIDE #87

Drexler's Cosmic-Ray Cosmology Applied to Galaxy Formation
Some Plausible Speculations

- When a UHE proton moving along a certain path crosses orthogonal magnetic field lines, it is deflected up or down depending upon the magnetic field direction. The degree of deflection is proportional to the orthogonal magnetic field strength.

- Any magnetic deflection of a UHE proton reduces its velocity in the direction of its original path for two reasons: the deflection itself will cause its direction to change, and synchrotron radiation losses will reduce its forward velocity.

- Thus, if UHE protons that are moving through space at a certain velocity encounter a bulge (increase) in the magnetic field strength, they will not pass through the magnetic bulge region as quickly as through a no-bulge region and could possibly linger in the region. To describe this magnetic attraction effect for relativistic protons and electrons, I will use the term "attract" in quotes.

SLIDE #88

Drexler's Cosmic-Ray Cosmology Applied to Galaxy Formation

Some Plausible Speculations

- Since electrons lose energy through synchrotron radiation at a rate 11 trillion times faster than protons, electrons would lose their energy quickly and tend to circulate and accumulate in magnetic-bulge regions.

- Such electron-filled regions would add a coulomb attractive force to UHE protons, slowing them down and also facilitating their conversion into hydrogen atoms. Conceivably, this process could lead, in some cases, to negatively charged proto-galaxies or galaxies surrounded by positively-charged UHE proton dark matter. This, in turn, could lead to lightning-like proton electrical discharges creating a multitude of gamma-ray bursts (GRBs) of immense proportions.

- As the UHE protons slow down in close proximity to a magnetic-field bulge they could linger close by, adding mass and gravitational attraction to the magnetic bulge region.

SLIDE #89

Drexler's Cosmic-Ray Cosmology Applied to Galaxy Formation

Some Plausible Speculations

- Magnetic-field bulges "attract" relativistic electrons and protons. The relativistic proton and electron "attraction" to a magnetic bulge may be different from, but as real as, the gravitational attraction between two masses.

- Previously, my concept of the "death spiral" was discussed with regard to UHE protons plummeting into the galaxy from the halo. In that case the galaxy magnetic field was about 2,000 times the strength of the extragalactic field. The Milky Way is an example of a significant magnetic-field bulge covering a large area in space that should capture almost all UHE protons spiraling through. In the solar system, the Sun is an example of a very significant magnetic-field bulge. With its magnetic field strength of about 50 gauss and the Milky Way's magnetic field strength at 2×10^{-6} gauss, the magnetic field ratio is 25 million.

SLIDE #90

Drexler's Cosmic-Ray Cosmology Applied to Galaxy Formation

Some Plausible Speculations

- It is widely believed that a fraction of a second after the Big Bang, the average energy of the UHE protons being dispersed was 10^{25} eV. This was a very efficient system for distributing mass, since each of those relativistic protons contained the equivalent mass of 10^{16} protons at rest.

- Following the Big Bang, weak magnetic fields were evolving in the Universe – probably created by UHE protons, electrons, and helium nuclei moving through space generating electric currents and therefore magnetic fields.

- Arrays of these primordial magnetic bulges might be better described as primordial magnetic ripples which could be the foundation for forming proto-galaxies.

SLIDE #91

Drexler's Cosmic-Ray Cosmology Applied to Galaxy Formation

Some Plausible Speculations

- When magnetic bulges have been created, the UHE protons spiraling through would incur energy losses from synchrotron radiation. The magnetic field of the bulge may strengthen and/or enlarge, leading to the formation of proto-galaxies. They then could capture most UHE protons spiraling through, thus creating concentrations of mass in one region and the absence of mass in an adjacent region.

- One of the mysteries of cosmology is the "large scale structure of the Universe" with its wide empty spaces and crowded groupings of galaxies and clusters and the famous Great Wall of galaxies. Perhaps the previous paragraph provides a possible explanation of how this structure could have come about.

- A study of the "large scale structures of the Universe" may lead to the conclusion that they can be more easily formed by the "attraction" of UHE protons and electrons to magnetic-field bulges in conjunction with gravitational attraction as compared with such formation by dark matter candidates proposed by others.

SLIDE #92

Drexler's Cosmic-Ray Cosmology Applied to Galaxy Formation

Some Plausible Speculations

- Since the General Theory of Relativity addresses only space, time, and gravitational field and mass, it may be inadequate to handle analyses of the Universe where most of the mass objects contain electric charges moving at relativistic velocities, which are "attracted" to magnetic-field bulges.

- It seems that the foregoing suggests that computer simulations of galaxy formation utilizing cold dark matter and warm dark matter should be re-examined. Perhaps they should include magnetic-field bulges and relativistic protons and electrons.

SLIDE #93

PART XIV

THE PRINCIPAL GOAL OF THESE LECTURES IS TO PROVIDE EVIDENCE

The Principal Goal of These Lectures Is to Provide Evidence

- Evidence that dark matter comprises a large number of ultra-high-energy protons moving through magnetic fields, which causes them to spiral around in halos of galaxies and galaxy clusters.

- Evidence that the miniscule extragalactic magnetic fields and the tiny magnetic fields of galaxies may be small, but that they control the spiral paths of the powerful and multitudinous UHE protons.

- Evidence that the accelerating expansion of the Universe, attributed to dark energy, probably comes about by the continual loss of relativistic mass of the dark matter halos surrounding galaxy clusters. Such a reduction in the dark matter halo mass around galaxy clusters would raise the galaxy clusters' velocities under the Law of Conservation of Linear Momentum and reduce each galaxy cluster's gravitational attraction to nearby galaxy clusters, thereby facilitating their more rapid separation.

SLIDE #95

The Principal Goal of These Lectures Is to Provide Evidence

- Evidence that the Sun may have been created by UHE protons leaving the dark matter halo around the Milky Way to plunge into the Milky Way as cosmic-ray protons in a synchrotron radiation "death spiral." Some cosmic-ray protons entered the solar system region and were captured by the Sun through gravitational field attraction and magnetic field "attraction."

- Evidence that the nature of dark matter, the accelerating expansion of the Universe (dark energy), and star formation are based upon a unified theory which involves UHE protons traveling through at least two levels of magnetic field strengths. (Note again that the coulomb charges of the UHE protons, the miniscule magnetic fields, and the spiral paths of the relativistic protons are not taken into account in Einstein's General Theory of Relativity.)

PART XV

DREXLER'S TENTATIVE CONCLUSIONS REGARDING DARK MATTER, ACCELERATING EXPANSION, STAR AND SUN FORMATION, GALAXY FORMATION, AND GENERAL ASTROPHYSICS THEORY

SLIDE #97

Tentative Conclusions:

Dark Matter

- Most cosmic-ray protons with energies at or above 10^{15} eV, showering the Earth, are probably relics of the Big Bang.

- The cosmic-ray protons showering the Earth have been studied for almost 90 years but have not been given their rightful place in the field of cosmology.

- The dark matter particles in halos around spiral galaxies and galaxy clusters today probably are UHE protons that are relics of the Big Bang 10^{25} eV protons which have lost all but perhaps a billionth or so of their original energy (and mass). Their typical kinetic energies are probably in the range of 10^{16} eV to 10^{17} eV.

SLIDE #98

Tentative Conclusions:

Dark Matter

- The size of the Milky Way galaxy* is consistent in size with the diameter of the spiral orbit of a 10^{16} eV proton moving through the 10^{-9} gauss extragalactic magnetic field.

- The size of the Milky Way galaxy's dark matter halo* is consistent with the size of the spiral orbit of a 10^{17} eV proton moving through the 10^{-9} gauss extragalactic magnetic field.

- If these UHE protons have determined both the size of galaxy and the size of the dark matter halo, this is strong evidence that UHE protons may be a principal constituent of dark matter.

- The widely searched-for theoretical neutralino particles are similar in certain respects to UHE protons, whose mass is in the right range and whose Earth penetration power may be similar to that expected of neutralinos.

*See slides #17, #85, and #86.

Tentative Conclusions:

Accelerating Expansion of the Universe Between Galaxy Clusters

- The accelerating expansion of the Universe may be caused by the declining mass of dark matter halo protons around galaxy clusters, moving through the extragalactic magnetic fields and losing energy (and mass) through synchrotron radiation.

- Synchrotron radiation losses of UHE protons in magnetic fields and their resulting mass decline and the Law of Conservation of Linear Momentum appear to be key factors in causing the accelerating expansion of the Universe.

- The "death spiral" transition of the halo UHE protons into cosmic-ray protons could lead to both the accelerating expansion of the Universe and to new star formation through accretion of those decelerated cosmic-ray protons.

Tentative Conclusions:

Star and Sun Formation

- UHE cosmic-ray protons showering the extended solar system region for billions of years probably were instrumental in creating the Sun and the solar system.

- The mass of the Sun and some other Population I stars in the Milky Way may be increasing as a result of cosmic-ray particle showers from the dark matter halo surrounding the Milky Way.

- The "death spiral" transition of the dark matter halo UHE protons into galaxy cosmic-ray protons probably led to accretion of cosmic-ray protons by the developing Sun through both gravitational field attraction and magnetic field "attraction."

Tentative Conclusions:

Galaxy Formation

- Magnetic field strength ripples in the primordial space may have been as significant in creating proto-galaxies as ripples in the gravitational field strength.

- Computer simulations of the formation of galaxies and the "large scale structure of the Universe" should include magnetic-field bulges and relativistic protons and electrons.

- It is conceivable that some spiral galaxies are negatively charged, surrounded by positively charged relativistic UHE protons. This could lead to lightning-like proton electrical discharges creating a multitude of gamma-ray bursts (GRBs) of immense proportions. (See slide #89.)

Tentative Conclusions:

General Astrophysics Theory

- The cosmic-ray protons showering the Earth today with energies of 10^{15} eV to 10^{20} eV are probably relics of the 10^{25} eV Big Bang protons. There are no other known proton sources that could provide these UHE proton energies or the enormous quantities of UHE protons. (See slides #33 - #35 and #46 - #48.)

- The vast majority of the UHE proton energy decline to one billionth of Big Bang levels is probably caused by synchrotron radiation losses. (See slides #38 - #44.)

- GZK cutoff losses are not considered to be material. (See slides #34 and #35.)

- Red-shifting proton energy losses were probably low in comparison to synchrotron radiation losses since the UHE protons would have been locked into spiral paths in the extragalactic magnetic fields. Therefore, their kinetic energy levels would not be affected significantly by the expansion of the Universe. (See slide #85.)

SLIDE #103

Tentative Conclusions:

General Astrophysics Theory

- The General Theory of Relativity, which is based upon space, time, gravitational fields, and mass, by itself should be incapable of complete solutions when the vast majority of particles in the Universe are positively-charged UHE protons and electrons moving through the fabric of space containing ripples in magnetic field intensity.

- The recently contested 1966 GZK cut-off theory must be relegated to a secondary status. It has caused many astrophysicists to not realize that UHE cosmic-ray protons may be a relic of the Big Bang.

- Big Bang theories do not point out the elegant efficiency of the creation of the Universe where protons with a kinetic energy of 10^{25} eV evolved from the Big Bang, each of which actually represents an equivalent mass of 10^{16} protons at rest.

SLIDE #104

Tentative Conclusions:

General Astrophysics Theory

- Particle synthesis theory during the Big Bang period does not seem to take into consideration the relativistic mass of the 10^{25} eV protons. Otherwise, the inflation-theory cosmologists might have used the relativistic mass of the protons to determine whether the mass density of the Universe would lead to the critical density or to an open or Closed Universe. (Note that at the same temperature for protons and helium nuclei, a UHE proton's relativistic mass can be orders of magnitude greater than that of helium nuclei. Thus, relativistic-mass abundances could be very different from rest-mass abundances. See slide #18.)

- Baryonic dark matter seems to fit the particle mass requirements of the inflationary cosmologists if the protons are moving at relativistic velocities.

SLIDE #105

IN CONCLUSION

Cosmic-ray cosmology is based both upon Albert Einstein's General Relativity equations and James Clerk Maxwell's electromagnetic/electrodynamic equations. The effects of coulomb charges moving at relativistic velocities through magnetic fields in the Universe may be too significant and extensive to be ignored.

Cosmic-ray cosmology represents a paradigm shift and a contrarian approach to some aspects of today's astrophysics and cosmology. A dozen of my principal ideas on this subject are briefly summarized in the appendix on the following two pages.

These lectures and this book are offered as a thought-provoking exploration and analysis. I hope that cosmic-ray cosmology will become a subject of discussion, critique, and debate.

Copyright © 2003 Jerome Drexler, Los Altos Hills, California.
All rights reserved.
Filed with the U.S. Patent and Trademark Office.

APPENDIX

A Dozen Contrarian or Paradigm-Shifting Astrophysics Ideas

(1) Relativistic protons moving through the extragalactic magnetic field of the Universe can satisfy the astrophysics requirements for dark matter.

(2) Dark matter is comprised of relativistic UHE protons circulating in the dark matter halos of spiral galaxies where they are trapped by the surrounding extragalactic magnetic field.

(3) The high gravitational field strength of the dark matter halos around spiral galaxies is derived from the relativistic masses of the trapped, circulating UHE protons.

(4) Particle abundances created in conjunction with the Big Bang should be analyzed from both rest-mass ratios and relativistic-mass ratios.

(5) The highest energy cosmic-ray protons are relics of the 10^{25} electron volt protons generated by the Big Bang.

(6) Only a portion of the then-existing protons combined with electrons to form hydrogen atoms 700,000 years after the Big Bang.

APPENDIX

A Dozen Contrarian or Paradigm-Shifting Astrophysics Ideas

(7) Cosmic-ray protons are derived from halo UHE protons that have left the dark matter halo and entered into the galaxy in "death spirals."

(8) The Sun was created by cosmic-ray protons and helium nuclei showering on the solar system extending to the Oort Cloud region.

(9) The Sun is more massive today than at the time of its birth.

(10) The mass of dark matter halos surrounding spiral galaxies continues to decline as a result of synchrotron radiation losses (and the formation of new stars).

(11) If the mass contained in the dark matter halos surrounding all galaxy clusters is declining, then the separation velocities between galaxy clusters should be increasing.

(12) Magnetic-field-strength bulges in the fabric of space will "attract" relativistic protons and electrons.

BIBLIOGRAPHY AND SUGGESTED SOURCES

J. N. Bahcall and J. P. Ostriker, *Unsolved Problems in Astrophysics* (Princeton Univ. Press, Princeton, New Jersey, 1997).

R. Clay and B. Dawson, *Cosmic Bullets – High Energy Particles in Astrophysics* (Helix Books, Addison-Wesley, Australia, 1998).

K. Croswell, *The Alchemy of the Heavens* (Anchor Books, Doubleday, New York, 1995).

D. Filkin, *Stephen Hawking's Universe* (Basic Books, New York, 1997).

M. W. Friedlander, *Cosmic Rays* (Harvard Univ. Press, Cambridge, MA and London, England, 1989).

M. W. Friedlander, *A Thin Cosmic Rain – Particles From Outer Space* (Harvard Univ. Press, Cambridge, MA and London, England, 2000).

H. Friedman, *The Astronomer's Universe* (W.W. Norton & Company, New York, 1998).

D. Goldsmith, *The Astronomers* (St. Martin's Press, New York, 1991).

A. H. Guth, *The Inflationary Universe* (Helix Books, Perseus Books, Reading, Massachusetts, 1998).

R. P. Kirshner, *The Extravagant Universe – Exploding Stars, Dark Energy and the Accelerating Cosmos* (Princeton Univ. Press, Princeton, New Jersey, 2002).

E. W. Kolb and M. S. Turner, *The Early Universe* (Addison-Wesley, USA, 1990).

M. S. Longair, *High Energy Astrophysics*, Volume I, Second Edition (Cambridge Univ. Press, Cambridge, UK 1992).

M. S. Longair, *High Energy Astrophysics*, Volume II, Second Edition (Cambridge Univ. Press, Cambridge, UK, 1994).

M. S. Madsen, *The Dynamic Cosmos* (Chapman & Hall, London, England, 1995).

V. Rubin, *Bright Galaxies – Dark Matters* (Amer. Inst. Physics, New York, 1997).

D. W. Sciama, *Modern Cosmology and the Dark Matter Problem* (Cambridge Univ. Press, Cambridge, UK, 1993).

J. Silk, *A Short History of the Universe* (Scientific American Library, a Division of HPHLP, New York, 1997).

T. X. Trinh, *The Secret Melody* (Oxford Univ. Press, New York, 1995).

J. N. Wilford, *Cosmic Dispatches – The New York Times Reports on Astronomy and Cosmology* (W. W. Norton & Company, New York and London 2002).

W. S. C. Williams, *Nuclear and Particle Physics* (Oxford Univ. Press, Oxford and New York, 1991).

GLOSSARY

ACCESS: Advanced Cosmic Ray Composition Experiment for the Space Station. The ACCESS experiment was designed to investigate the "knee" of the cosmic-ray energy distribution graph.

Accretion: An infall of matter on an object.

AGASA: Akeno Giant Air Shower Array.

Alpha particle: A particle consisting of two protons and two neutrons bound together. The nucleus of a helium atom is an alpha particle.

Andromeda: Twin galaxy of the Milky Way. Together the two galaxies comprise most of the mass of the Local Group.

ASCA: Advanced Satellite for Cosmology and Astrophysics was on an X ray mission 1993-2001.

Astronomical unit (A.U.): The average distance from the Sun to the Earth, equal to 149,598,000 kilometers.

Astrophysics: The study of the composition and other physical properties of celestial objects.

"Attract"/"attraction": A new term (in quotes) that refers to the movement in space of a UHE relativistic proton from one magnetic field strength to a higher magnetic field strength, resulting in the slowing down and dwelling in the region of the higher magnetic field.

Baryon/baryonic: An elementary particle that is subject to the strong nuclear interaction. The proton and neutron are baryons.

Beryllium: A steel-gray, light, strong, brittle, toxic, bivalent, metallic element.

Big Bang: The cosmological theory that holds that all the matter and energy in the Universe was concentrated in an immensely hot and dense point, which exploded 13.7 billion years ago.

Black hole: An object that exerts such enormous gravitational force that nothing, not even light or other forms of electromagnetic radiation, can escape from it.

Boron: A trivalent, metalloid element found in nature.

CERN: Center for European Nuclear Research.

Closed Universe: A universe in which the density of matter is greater than the critical density and that will thus collapse onto itself in the future.

COBE: Cosmic Background Explorer.

"Cold" dark matter: Non-baryonic matter consisting of elementary particles of relatively high mass that are moving relatively slowly. (The term "cold" indicates a low temperature and thus a small energy of motion.)

Collision cross section: A measure of the probability that an encounter between particles will result in the occurrence of a particular atomic or nuclear reaction.

Coma cluster: A galaxy cluster that contains about 1,000 galaxies. The gravitational effects of dark matter were discovered in this galaxy cluster.

Comet: A body of ice and dust, with a nucleus of typically about 10 kilometers in diameter. It is visible only when it travels close enough to the Sun to reflect light.

Cosmic background radiation (CBR): The microwave radiation that bathes the entire Universe and that dates from the epoch when the Universe was just 300,000 years old.

Cosmic-ray cosmology: A new term to describe a cosmology recently developed by J. Drexler, based upon UHE protons and cosmic-ray protons.

Cosmic rays: Particles (mostly protons and electrons) that have been accelerated somewhere in the Universe to very high energies.

Cosmology: The study of the Universe as a whole, and of its structure and evolution.

Coulomb force: The force between two coulomb charges or electrically charged particles.

Dark energy (as defined in the past): A hypothetical form of energy that permeates all space and has negative pressure resulting in a repulsive gravitational force. The accelerating expansion of the Universe has been attributed to dark energy.

Dark matter (as defined in the past): Matter that is detected only by its gravitational pull on visible matter. The composition has been unknown; it might consist of very low mass stars or supermassive black holes, but big-bang nucleosynthesis calculations limit the amount of such baryonic matter to a small fraction of the critical mass density. If the mass density is critical, as predicted by the simplest versions of inflation, then the bulk of the dark matter must be a gas of weakly interacting non-baryonic particles, sometimes called WIMPS (Weakly Interacting Massive Particles).

"Death spiral": A new term (in quotes) that refers to the spiral path of a UHE relativistic proton moving from a lower magnetic field to a considerably higher magnetic field wherein the radius of curvature of the UHE proton's spiral path is greatly reduced, leading to a significant increase in the synchrotron radiation losses and rapid decrease of the proton's kinetic energy.

Deuterium: A chemical element whose nucleus consists of a proton and a neutron, created mainly in the first three minutes of the Universe's history.

Doppler effect: The variation in the energy and color of light caused by the motion of a source of light relative to an observer. If the source is receding, the energy decreases and the light is shifted toward the red. If the source is approaching, the energy increases and the light is shifted toward the blue.

Doppler shift: The shift in the received frequency and wavelength of an electromagnetic wave that occurs when either the source or the observer is in motion. Approach causes a shift toward shorter wavelengths and higher frequencies called a blue shift. Recession has the opposite effect, called a red shift. The expansion of the Universe causes ancient electromagnetic wave emissions to exhibit a doppler red shift.

Electromagnetic wave: A pattern of electric and magnetic fields that moves through space. Depending on the wavelength, an electromagnetic wave can be a radio wave, a microwave, an infrared wave, a wave of visible light, an ultraviolet wave, a beam of X rays, or a beam of gamma rays.

Electron: The lightest of the subatomic particles with electrical charge. The electron has a mass of 9×10^{-28} kilograms and is negatively charged.

Electron volt (eV): The energy released when a single electron passes through a one-volt battery.

Elliptical galaxy: A galaxy observed as an oval-shaped system generally composed of old stars, a large black hole, and containing little or no gas and dust.

Extragalactic: The regions of the Universe outside of any galaxy.

Galactic disk: A flattened aggregation of stars, gas, and dust in a spiral galaxy. The average disk is some 90,000 light-years in diameter and 300 light-years thick. In the Milky Way, the stars complete one turn around the galactic center every 250 million years, at a velocity of 230 kilometers per second.

Galactic halo: A spherical region around a spiral galaxy populated by old stars and globular clusters. Observations suggest that it is surrounded by an invisible (dark matter) halo some 10 to 20 times larger and more massive.

Galaxy: A system of stars (10 million in a dwarf galaxy, 100 billion in an average galaxy like the Milky Way, 10 trillion in a giant galaxy) held together by gravity.

Galaxy cluster: A dense grouping of several thousand galaxies bound by gravity, with an average diameter of some 60 million light-years, and an average mass of a few million billion solar masses.

Gamma ray: An electromagnetic wave with a wavelength in the range of 10^{-13} to 10^{-10} meters, corresponding to photons with energy in the range of 10^4 to 10^7 electron volts. Their energies are higher than X rays.

Gauss: A measure of the strength of a magnetic field.

General relativity: A gravitational theory proposed by Albert Einstein in 1915, which is more accurate than that of Newton. The two theories differ mainly in situations where gravitational fields are very intense, such as around a pulsar or black hole. General relativity constitutes the theoretical support of the Big Bang theory.

GeV: G stands for Giga, or 10^9. Thus, GeV is one billion electron volts.

Gravitational field: A field of force surrounding a body of finite mass. The field of force is defined as the force that would be experienced by a standard mass positioned at each point in the field.

Gravitational force: The force responsible for attraction between all matter. The weakest of the four forces, it also possesses the longest range.

GRB: A gamma ray burst.

Great Attractor: A large grouping of galaxies with a total mass of 100 million billion solar masses, gravitationally attracting the Local Supercluster, which is moving toward it.

Great Wall: A sheet of galaxies which stretches more than 500 million light-years across the sky.

Group of galaxies: A collection of about 20 galaxies held together by gravity, some six million light-years across and averaging between one and 10 trillion solar masses.

GZK Cosmic-Ray cutoff: A theory limiting proton energies. According to the currently questioned 1966 Greisen-Zatsepin-Kuzmin (GZK) cutoff theory, protons with energies greater than 6×10^{19} eV would interact with the cosmic microwave background radiation and lose energy through radiation and thus would not travel more then 50 Mpc, or about 160 million light-years. In 1998 Coleman and Glashow wrote a paper entitled, "Evading the GZK Cosmic-Ray Cutoff" which showed that for very high energy cosmic rays, the GZK cutoff would not apply.

Halo: The region around a galaxy that contains dark matter and some stars.

Heavy elements: All chemical elements with nuclei heavier than helium. Also known as "metals," these heavy elements are built up by nuclear fusion in the interiors of stars and supernovae.

Helium: A chemical element with a nucleus of two protons and two neutrons (helium-4). A second, far-less-abundant isotope has two protons and one neutron (helium-3).

Hydrogen: The lightest of all chemical elements, consisting of one proton and one electron. Hydrogen makes up 75% of the mass of the Universe.

Isotropy/isotropic: The property of the Universe to be similar in every direction.

Kpc: The abbreviation for a kilo parsec where a parsec equals 3.26 light-years.

Larmor Radius (for a proton): A proton crossing an orthogonal/magnetic field and entering into a spiral path. The radius of a cycle of that spiral path is called the proton Larmor Radius for that cycle.

$$\text{Proton Larmor Radius} = 110 \text{ Kpc} \times \frac{10^{-8}}{B} \text{ gauss} \times \frac{E}{10^{18} \text{ eV}}$$

Light-year: The distance traveled by light (which moves at a velocity of 300,000 kilometers per second) in 1 year, and equal to 9,460 billion kilometers.

Local Group: A grouping of galaxies – extending over a region of space of about 10 million light-years – of which the Milky Way and Andromeda are the principal and most massive members (one trillion solar masses each). It also includes dwarf galaxies ranging from 10 million to 10 billion solar masses.

Local Supercluster: A loosely knit assemblage of some 100 clusters of galaxies, including the Local Group.

Magnetic field: A field of force in space, created by a magnet or by an electric current, that guides the trajectories of electrically charged particles by exerting an electromagnetic force.

Mass: The measure of the inertia of an object, determined by observing the acceleration when a known force is applied. An object with mass creates a gravitational field, which is defined in this glossary. When a proton travels at relativistic velocities, it has a relativistic mass equal to its energy divided by the square of the speed of light.

Maxwell's Equations: A set of differential equations describing space and time dependence of the electromagnetic field and forming the basis for classical electrodynamics.

Microwave: An electromagnetic wave with a wavelength of between one millimeter and 30 centimeters.

Milky Way: The galaxy to which our solar system belongs, whose central regions appear as a band of light or "milky way" that we can see from Earth in clear night skies.

Missing mass: An outmoded name for the dark matter in the Universe.

M_o: The symbol m_o representing the mass of a proton when it is not moving (the rest mass).

Momentum: The linear momentum of an object, equaling the product of its mass and velocity. If no external forces are acting on a group of mass objects, the Law of Conservation of Linear Momentum requires that the total linear momentum of the mass objects in the group remains unchanged.

Neutralino: A theoretical non-baryonic particle, which is an amalgam of the superpartners of the photon (which transmits the electromagnetic force), the Z boson (which transmits the so-called weak nuclear force), and perhaps other particle types. Although the neutralino is heavy by normal standards (at least 35 times the mass of a proton), it is generally thought to be the lightest supersymmetric particle.

Neutron: A subatomic particle with no electric charge, one of the two basic constituents of an atomic nucleus.

Open Universe: A Universe in which the density of matter is less than the critical density and which will thus expand forever.

Oort cloud: A region in the outer limits of the solar system where billions to trillions of comets reside.

Orthogonal: Intersecting or lying at right angles.

Parsec: An astronomical unit of distance equal to 3.26 light-years or approximately 19 trillion miles.

Population I stars: A younger generation of stars with ages from a few million years to about 10 billion years and with a relatively large fractional abundance (about 1% of mass) of elements heavier than helium. The Sun is in this category.

Primordial ripples: The mass perturbations in the early Universe that evolved into galaxies, stars, etc.

Proto-galaxy: A cloud of gas and ions that is evolving into a galaxy.

Proton: A positively charged particle composed of three quarks, which, together with the neutron, forms atomic nuclei. The proton is 1836 times more massive than the electron.

Recombination: In the traditional theory, between 300,000 and 700,000 years after the Big Bang, the plasma of free electrons and hydrogen nuclei that condensed to form a neutral gas, in a process called recombination. The prefix "re" is not meaningful here, however, since according to the Big Bang theory the electrons and protons (hydrogen nuclei) were combining for the first time ever.

Red shift: A shift to longer wavelengths and lower frequencies, typically caused by the doppler effect in a receding object or caused by the expansion of the Universe.

Solar system: The Sun and the objects in orbit around it, which include nine planets, nearly 60 known satellites of the planets, thousands of smaller objects called asteroids, and billions to trillions of comets.

Solar wind: Streams (up to one trillion) of charged particles expelled from the hottest, outermost layers of the sun, which pass the Earth and the other planets as bunches of electrons and ions, the latter typically nuclei of hydrogen, helium, oxygen, carbon, nitrogen, and neon that have lost one or more of the electrons that orbit each nucleus. Astronomers estimate that about 20% of the Sun's mass has been lost due to the solar wind.

Spiral galaxy: A flattened, disk-like system of stars and interstellar gas and dust with a spherical collection of stars, known as the bulge, at its center. Bright, young stars outline spiral arms in the plane of the disk.

(The) Standard Model: Name given to the current theory of fundamental particles and how they interact.

Star: A sphere of gas consisting of 98% hydrogen and helium and 2% heavy elements in equilibrium under the action of two opposing forces: the compressive gravity and the outward radiation pressure from the nuclear fusion reactions in its core. The Sun has a mass of 2×10^{30} kilograms, and masses of stars range between 0.1 and 100 solar masses.

Supercluster: The aggregation of tens of thousands of galaxies held together by gravity and gathered into groups and clusters. Superclusters have the shape of flattened pancakes with an average diameter of 90 million light-years and masses of 10,000 trillion (10^{16}) solar masses.

Supernova/supernovae: An exploding star, visible for weeks or months, even at enormous distances, because of the tremendous amounts of energy that the star produces. Supernovae typically arise when massive stars exhaust all means of producing energy from nuclear fusion. In these stars, the collapse of the star's core results in the explosion of the star's outer layers. Another type of supernova arises when hydrogen-rich matter from a companion star accumulates on the surface of a white dwarf and then undergoes nuclear fusion. This second type, known as a Type 1a supernova, generates light at a well-known standard level and thus can be used to measure the rate of expansion of the Universe.

Synchrotron radiation: Electromagnetic radiation that is emitted by charged particles moving at relativistic speeds in circular orbits in a magnetic field. The rate of emission is inversely proportional to the product of the radius of curvature of the orbit and the fourth power of the mass of the particles. For this reason, synchrotron radiation is not a problem in the design of proton synchrotrons but it is significant in electron synchrotrons. Synchrotron radiation in space usually is in the form of X rays or gamma rays.

TeV: T stands for Tera, or 10^{12}. Thus, TeV is one trillion electron volts.

UHE proton: A proton traveling near the speed of light. UHE stands for ultra high energy.

Ultraviolet (UV): Ultra violet light.

White dwarf star: A small, dense star with a diameter of about 10,000 kilometers (about the size of Earth) created when a star of less than 1.4 solar masses exhausts the nuclear fuel and collapses under its own gravity. This type of star participates in a Type 1a supernova.

WIMP: Weakly Interacting Massive Particle. The name for a non-baryonic theoretical dark matter candidate that is presumed to have a mass much greater than that of a proton.

X rays: Electromagnetic radiation with greater frequencies and smaller wavelengths than those of ultraviolet radiation and lower frequencies and longer wavelengths than those of gamma ray radiation.

INDEX

ACCESS, 24, 33, **119**

Accretion, 74, 75, 76, 83, 107, 108, **119**

Alpha particle, 2, **119**

Andromeda, 5, 65, **119**, 127

ASCA, 52, **119**

Astronomical unit (A.U.), 73, **119**, 129

Astrophysics, 114, **119**

"Attract"/"attracted"/"attraction", 95, 97, 99, 100, 103, 108, 115, **120**

Bahcall, John N., 47, 116

Baryon/baryonic/non-baryonic, iv, v, 1, 12, 13, 15, 17, 18, 20, 36, 37, 112, **120**, 121, 122, 129, 133

Bertsch, David, 49

Beryllium, v, 28, 37, **120**

Bhattacharjee, P., 55

Big Bang, **120**, *et al*

Black hole, 88, 90, **120**, 122, 124, 125

Boron, v, 28, 37, **120**

Cal Tech, 43

Case Western Reserve University, 60

CERN/*CERN Courier*, 24, 32, 33, **120**

Clay, Roger, 34, 116

Cline, David B., 14, 15, 16, 17, 20, 21, 25

Closed Universe, 112, **120**

COBE, 13, **120**

"Cold" dark matter, 15, **121**

Coleman, Sidney, 41, 42, 62, 126

Collision cross section, 78, **121**

Coma cluster, 9, **121**

Comet, 73, **121**, 129, 131

Cosmic background radiation (CBR), 41, 62, **121**

Cosmic ray, **121**, *et al*

Cosmic-ray cosmology, vii, viii, 79, 80, 81, 82, 83, 90, 91, 92, 93, 95, 96, 97, 98, 99, 100, 113, **121**

Cosmology, **122**, *et al*

Coulomb charge, 103, 113, 122

Coulomb force, 96, **122**

Cronin, James W., 47, 54

Dark energy, **122**, *et al*

Dark matter, **122**, *et al*

Dark matter halo, iv, v, vii, 3, 5, 7, 9, 11, 31, 32, 35, 37, 44, 66, 67, 68, 72, 80, 81, 82, 86, 88, 89, 90, 93, 102, 103, 106, 107, 108, 114, 115

Dawson, Bruce, 34, 116

"Death spiral", vii, 4, 73, 84, 85, 86, 87, 88, 97, 103, 107, 108, **122**, 115

Deuterium, **123**

Dixon, David D., 50

Doppler effect/doppler shift, **123**, 128

Einstein, 80, 81, 103, 113, 125

Electromagnetic wave, x, **123**, 125, 128

Electron, **123**, *et al*

Electron volt (eV), **124**, *et al*

Elliptical galaxy, 90, **124**

Enrico Fermi Institute, 55

Europhysics News, 41, 42, 62

Extragalactic, 3, 4, 7, 44, 47, 54, 64, 81, 86, 87, 92, 93, 94, 97, 102, 106, 107, 110, 114, **124**

Friedman, Herbert, 2, 54, 72, 116

Galactic disk, **124**

Galactic halo, **124**

Galaxy, **124**, *et al*

Galaxy cluster, **124**, *et al*

Gamma ray, vi, 46, 48, 49, 50, 55, 63, 66, 121, **125**, 132, 133

Gauss, 47, 73, 86, 92, 93, 94, 97, 106, **125**, 127

General relativity, 74, 80, 81, 100, 103, 111, 113, **125**

Gehrels, Neil, 49

GeV, 40, 51, **125**

Glashow, Sheldon L., 41, 42, 62, 126

Gravitational field, 35, 73, 83, 100, 103, 108, 109, 111, 114, **125**, 128

Gravitational force, 57, 58, 65, 78, 118, 120, **125**

GRB, 51, 96, 109, **125**

Great Attractor, **126**

Great Wall, 99, **126**

Group of galaxies, **126**, *et al*

Gupta, N., 55

Guth, Alan H., 13, 20, 40, 116

GZK, 41, 42, 62, 110, 111, **126**

Halo, **126**, *et al*

Hartman, Robert, 49

Harvard-Smithsonian, 59

Heavy elements, 28, 72, **126**, 131

Helium, **127**, *et al*

Hillas, A.M., 33

Hubble, Edwin P., 4, 5, 57

Hydrogen, **127**, *et al*

Isotropy/isotropic, **127**

Kings University College, 40

Kirshner, Robert P., 59, 60, 117

Knee, 24, 33, 35, 37, 94, 119

Kosowsky, A., 43

Kpc, 92, 93, 94, **127**

Larmor Radius, viii, 3, 46, 86, 87, 92, 93, 94, **127**

Lawrence Berkeley National Laboratory, 57

Light-year, 42, 50, 59, 92, 93, 121, 124, 126, **127**, 129, 132

Linear momentum/momentum, 6, 7, 66, 67, 68, 82, 102, 107, **129**

Local Group, 4, 36, 37, 65, 117, **127**, 128

Local Supercluster, 54, 65, 126, **128**

Longair, M.S., 54, 117

Macomb, Daryl, 49

Magnetic field, **128**, *et al*

Mass, **128**, et al

Massachusetts Institute of Technology (MIT), 13

Maxwell, 113, **128**

Microwave, x, 42, 121, 123, 126, **128**

Milky Way, **128**, *et al*

Missing mass, 50, **128**

M_o, **129**, *et al*

Momentum/linear momentum, 6, 7, 66, 67, 68, 82, 102, 107, **129**

NASA, 29, 30, 49

Nature, 36, 37, 49

Neutralino, 1, 15, 16, 106, **129**

Neutron, 1, 12, 14, 20, 119, 120, 123, 127, **129**, 130

Olinto, A.V., 55

Oort cloud, 73, 74, 115, **129**

Open Universe, **129**

Orthogonal, 46, 92, 95, 127, **129**

Ostriker, Jeremiah P., 47, 116

Parsec, 42, 92, 127, **129**

Perlmutter, Saul, 57

Population I stars, vii, 71, 76, 83, 108, **130**

Primordial ripples, 13, 98, 109, **130**

Proto-galaxy, 96, 98, 99, 109, **130**

Proton, **130**, *et al*

Quasar, 36

Recombination, **130**

Red shift, 43, 110, 123, **130**

Riess, Adam, 60, 65

Rubin, Vera, 9, 117

Scientific American, 15, 16, 17, 18

Solar system, **131**, *et al*

Solar wind, 75, 77, **131**

Space Telescope Science Institute, 60

Spiral galaxy, iv, 11, 28, 85, 87, 93, 114, 115, 124, **131**

(The) Standard Model, 15, **131**

Star, **131**, *et al*

Sun, iii, vi, vii, viii, ix, 2, 3, 7, 8, 59, 69, 70, 71, 72, 73, 74, 75, 76, 77, 78, 83, 85, 97, 103, 104, 108, 115, 119, 121, 130, 131

Supercluster, 54, 65, 122, 124, **132**

Supernova/supernovae, 5, 18, 57, 59, 126, **132**, 133

Synchrotron radiation, **132**, *et al*

TeV, 51, 52, 55, **133**

Theory of Relativity, 74, 80, 81, 100, 103, 111, 113, **125**

Thin-disk stars, 70, 71, 76

Thompson, David, 49

Tonry, John, 60

Totani, T., 51

Trinh, Thuan Xuan, 61, 117

Turner, Michael S., 61, 65, 117

UHE proton, **133**, *et al*

Ultraviolet (UV), 36, 122, **133**

University of Birmingham (U.K.), 29

University of California, Berkeley, 43,

University of California, Riverside, 50

University of Chicago, 55, 61

University of Tennessee, 43

U.S. Department of Energy, 77

Watson, A. A., 33

White dwarf, 130, **132**

Wilford, John Noble, 58, 118

WIMP, 1, 120, **133**

X rays, vi, 46, 52, 119, 123, 125, 132, **133**

Zwicky, Fritz, 9

Printed in the United States
33576LVS00003B/169